LEADING
THROUGH
TURBULENCE

LEADING THROUGH TURBULENCE

How a Values-Based Culture Can Build Profits and Make the World a Better Place

ALAN LEWIS AND HARRIET LEWIS

NEW YORK   CHICAGO   SAN FRANCISCO
LISBON   LONDON   MADRID   MEXICO CITY   MILAN
NEW DELHI   SAN JUAN   SEOUL   SINGAPORE
SYDNEY   TORONTO

The **McGraw·Hill** *Companies*

1 2 3 4 5 6 7 8 9 10   DOC/DOC   1 8 7 6 5 4 3 2

ISBN 978-0-07-177710-0
MHID   0-07-177710-5

e-ISBN 978-0-07-177709-4
e-MHID   0-07-177709-1

Design by Lee Fukui and Mauna Eichner

McGraw-Hill products are available at special quantity discounts to use as premiums and sales promotions or for use in corporate training programs. To contact a representative, please e-mail us at bulksales@mcgraw-hill.com.

This book is printed on acid-free paper.

To our children, Edward and Charlotte, with love

Contents

Foreword   ix

CHAPTER 1
Developing Building Blocks for Success   1

CHAPTER 2
Achieving Success by Adhering to Mission,
Vision, and Values   29

CHAPTER 3
Leadership from Anywhere (and Everywhere)   51

CHAPTER 4
Doing Well by Doing Good:
Integrating Philanthropy with Business Strategy   77

CHAPTER 5
Delivering Unsurpassed Value   101

CHAPTER 6
Measuring for Excellence (Again and Again)   119

CHAPTER 7
Fostering Customer Loyalty and Building Community   135

CHAPTER 8
Creating Marketing Strategies to
Leverage Extreme Competitive Advantages   145

CHAPTER 9
Thriving in Change   161

CHAPTER 10
Managing Crises: Moments of Truth   193

APPENDIX A
Where We Give Back, 2011   221

APPENDIX B
Recognition of Our Commitment to Give Back   227

APPENDIX C
Sample Customer Survey   229

Index   231

Foreword

In 1985, Grand Circle Travel was a small travel company with $27 million in sales that was losing $2 million a year. Then entrepreneurs Alan and Harriet Lewis acquired it. Today Grand Circle is one of the largest and most successful tour operators in the world with $500 million in annual sales, more than 115,000 travelers each year, three travel brands, and a worldwide organization that is headquartered in Boston with 34 overseas offices and 2,200 people supporting it from 60 countries. Grand Circle's charitable foundation has donated or pledged $91 million to more than 300 educational, humanitarian, and cultural causes around the world, including 100 schools in 60 communities in 30 countries.

Few companies in this business have managed to grow to this scale in the volatile environment of international travel. Grand Circle's secret? Over the years, the company has created an adaptive leadership model that has helped it to not only survive but even thrive in a hostile, constantly changing environment. This unconventional model might prove very effective for other organizations trying to grow in or adapt to unpredictable times.

The Five Big Questions This Book Answers

1. In an unpredictable world, why is a different model needed from the traditional model taught in business schools?

2. How can I build an organization that can adapt and thrive in change and crisis?

3. How can a systematic focus on corporate culture and values create an organization that retains exceptional talent and has the flexibility to adapt to any situation?

4. How does an investment in leadership at all levels of an organization pay off?

5. How can strategic corporate philanthropy help a company do well by doing good?

In this book, Alan and Harriet Lewis share the story of Grand Circle and of how they realized their dream of creating a very different kind of company. In doing so, they share key lessons they've learned along the way, from the risks they've taken, to the mistakes they've made, to the challenges they've overcome, as they built one of the most innovative and adaptive organizations in the world today.

—Leonard A. Schlesinger
President, Babson College

LEADING THROUGH TURBULENCE

Developing Building Blocks for Success

Do not go where the path may lead. Go instead where there is no path and leave a trail.

—RALPH WALDO EMERSON

Alan: In 1985, Grand Circle Travel (GCT) was a small travel company with $27 million in sales—and it was losing $2 million a year.

Twenty-six years later, Grand Circle is one of the largest and most successful international tour operators in the world. More than 115,000 people travel with us every year; that's more than 2 million people over the past quarter century. More than 2,200 people make up our team—working in our Boston headquarters, in one of our 34 offices overseas, or in one of some 60 countries where we operate our trips.

We have three travel brands now: Grand Circle Travel, Overseas Adventure Travel (OAT), and Grand Circle Cruise Line, a fleet of more than 50 small ships that we own and operate to serve our Grand Circle Travel and Overseas Adventure Travel brands. We're highly profitable, with annual sales of more than $500 million. And we're philanthropic, having donated or pledged more than $91 million to more than 300 educational, humanitarian, and cultural organizations worldwide through our charitable foundation, including 100 schools in 60 communities in 30 countries.

In the volatile environment of international travel, few companies have managed to thrive amid the many crises we've faced and grown to the scale that we have. The travel industry isn't the only volatile business, of course; the world is an increasingly competitive and unpredictable place. Many businesses today across a wide swath of industries are searching for a different model from the traditional business model taught in business schools. Neither Harriet nor I went to business school, and we built our hugely successful business in our own unconventional way. We believe the story of how our company has evolved and the lessons we've learned can help other businesses in other industries to grow or adapt in unpredictable times. That's why we've written this book.

How We Got Started

Harriet: Alan and I didn't set out to become entrepreneurs. I was raised in a traditional New England family, went to college, and became a special education teacher. Alan was a rebellious and street-savvy kid who ran with some rough company in high school. He tried college but was too restless for lecture halls and libraries, and so he quit after a year and moved from Boston to Florida, where he took a job as a beach lifeguard.

Teacher and lifeguard—there's no business school in either of those résumés, and it wasn't easy getting to where we are now, running a company that makes $500 million in revenues a year. Alan and I are self-made millionaires, our business is one of a kind, and we believe the company we built is pretty fabulous. We also believe that our model might be useful to other business leaders seeking to navigate an increasingly unpredictable world—and that our story might encourage other entrepreneurs to pursue their own dreams. I guess the best way to begin is at the beginning, and so here are some lessons we learned when we were first getting started.

Be Open to Opportunity

Alan: I got into the travel business because of my father, who was a principal of a small, Boston-based travel company called United Travel Service. My parents had divorced when I was very young, and I didn't see my father much when I was growing up. But he came to visit me while I was working as a lifeguard in Florida. I was feeling pretty good about myself because I had recently saved a man's life, and the story had made the newspapers. But my father couldn't care less about my new "hero" status; instead, he was concerned about the direction of my life—or rather, the lack of direction. He didn't see much future in lifeguarding, and so he offered me a job in his travel business. I was angry and disappointed that he wasn't at all impressed with my lifesaving achievements, but after a couple of days, I realized the travel job sounded interesting. So I swallowed my pride and accepted his offer.

I never expected to be working for my father, and I never expected to be working in the travel business. I really didn't have any expectations at all at that time. I was still a kid in my late teens, and like a lot of kids, I didn't know what I wanted to do with my life. This was just an opportunity that presented itself to me, and I took advantage of it.

In my new job, I flew overseas to Majorca. The island brought my senses alive. I had never experienced a different place with an entirely different history from ours, or people who ate different foods, spoke different languages, and lived in houses that looked nothing like those in Boston. I loved it, and I instantly knew that the world would become my classroom. By 1970, I was working as a tour guide. I got to go on cruises, and I traveled to Switzerland, England, Mexico, and the Mediterranean. It was all pretty exciting for a 21-year-old, and my passion for travel has never abated, not even to this day, some 40 years later.

I've talked to many people over the years who have told me similar stories of how they got started in their line of work. They got a lead from someone, got a job, realized they liked what they were doing, and became increasingly successful. There was no brilliant interview, no obscure Ph.D. thesis, no lightning bolt of inspiration—just a good opportunity wisely taken. The important thing is to recognize that good opportunity when it comes your way. Don't squander it! Don't let pride, or uncertainty, or fear get in the way. Learn everything you can from anyone who offers to help you in business, and always be open to new experiences.

Follow Your Passion

Alan: I loved my job with my father's company; it gave me purpose and direction and a whole new perspective on the world. But I also knew I wanted to be my own boss; I've never been good with authority. So in 1973, I started my own travel company, Trans National Travel (TNT), with a partner, offering group-travel vacations all over the world. Starting my own business wasn't something I had planned to do; I just knew I didn't want to work for someone else, and I was sure I'd be successful. Call it the confidence of youth. Of course, at 24 years old, I didn't really have anything to lose!

I realize it's more difficult to start a business when you're older, because you don't have the crazy moxie that you have when you're young and unencumbered. It's also easier to take chances in middle age if you've been taking chances all your life. But you can't let age and obligations keep you from following your passion. If you have a dream, you've got to go for it. If that means negotiating a start-up strategy with your spouse, then sit down at the kitchen table and do it. If it means raiding the savings account, ask every member of the family to find ways to help out. If you have to work two jobs to cover your bets, then get cracking. Build your plan; prepare for it; do the groundwork required to realize your dream.

Yes, your venture might fail. If it does, you start over. Scary? Yes, again. But not as scary, to my mind, as a lifetime of regrets.

I was lucky with my first private venture. My partner and I built TNT into a very successful company, and we multiplied our original investment many times over. But I had even bigger dreams, and after 10 years I was ready to move on. I wanted to build a travel company that would help change other people's lives, too. By this time, Harriet and I had gotten married, and we had two children, Charlotte and Edward. They were only one and three years old, and so it was a big risk for us to leave this company and start over. We needed to do some serious thinking about our future. So in 1985, Harriet and I sold our interest in TNT, and we took an extended vacation to unwind and decide what to do next. Little did we know how quickly that next opportunity would come our way.

Don't Be Afraid to Dream Big

Harriet: We were two weeks into our monthlong seaside vacation on Captiva Island in Florida when Alan announced that a company called Grand Circle Travel was up for sale, and he wanted to buy it. That's Alan—restless, driving, always looking for the next big challenge. It didn't even surprise me that we had to make a decision right away because another company was expected to sign a contract for Grand Circle the very next day!

In 1985, Grand Circle Travel was a small travel company that had fallen a bit on hard times. It had long operated profitably as the travel service for the American Association of Retired Persons under the management of AARP's insurance provider, Colonial Penn. But recently, it had run into trouble. Although it had sales of $27 million and a good list of AARP travelers, Grand Circle was losing more than $2 million a year, and Colonial Penn was looking to unload it. Alan and I had talked about buying it, but Alan had a noncompete agreement with his TNT partner that prevented him from entering a directly competitive business for five years and from doing *anything* in the travel industry for at least two years. Grand Circle wasn't directly competitive with TNT, but it was certainly a travel business, and so we were in a tough spot.

While we were discussing all this, Alan heard from a friend in the business that Grand Circle had lined up another buyer, Saga Holidays, a big British travel company that had just opened operations on our home turf, Boston. If we wanted to buy Grand Circle, we would need to make an immediate offer, one that would trump Saga's bid. And we would have to figure out what to do about the noncompete.

Alan was ready to go, but I worried that a new venture would be hard on our family. On the other hand, Grand Circle looked like a great opportunity. Yes, it was losing money, but the company was a good size, it had a well-defined market and loyal customer base, and it had a serious commitment to travel. Travel has always been our passion. It has changed and enriched our lives in countless wonderful ways. Besides, we had built one successful business; why not another?

These were the considerations Alan and I discussed while walking on the beach on Captiva Island, as we tried to make the most momentous decision of our lives. We talked about how we wanted to give other Americans the opportunity to experience travel the way we had experienced it: up close and personal and with a deep human connection. We also wanted to build a company where people could look forward to coming to work, where they could grow into leadership roles and enrich their personal lives. In the end, the decision was easy: we realized Grand Circle was the path to realizing our dream of changing people's lives in our company, our community, and the world. With butterflies in our stomachs, we decided to go for it. Alan caught the next plane to New York to try to buy Grand Circle from Colonial Penn before Saga could close its deal.

Take Risks with a Clear Conscience

Harriet: On the plane north from Florida, Alan figured he was in for a fight. When he got to New York, he literally barged into the negotiations between Colonial Penn and Saga. The Saga people couldn't believe it. They were very eager to buy Grand Circle, hoping to turn that list of AARP travelers into brand-new Saga customers. They had fire in their bellies and dollar signs in their eyes. Then along comes this gate-crasher, brashly shaking hands and waving a checkbook.

Alan was maybe a little less confident than he looked, for he had to make an on-the-spot calculation. If we really wanted Grand Circle, we'd need to bid higher than Saga. But how much higher? What was the company worth to us? We already had enough money to live comfortably and travel for the rest of our lives. Was it worth risking it all? And what about the noncompete agreement? We didn't have time to get a legal opinion about what would happen if Alan's bid was accepted. Were we willing to risk an extended court fight?

Alan and I are risk takers by nature. We're not afraid of making mistakes and taking some losses. We also believe in decisiveness. We've seen too many businesspeople who become paralyzed at the point of decision making. They can't decide. They need more time. They need more information. They want a cost-benefit analysis. They want the perfect outcome. We seldom dither in this way, and we don't let our associates do it either. Make a decision, we tell them, *any decision*, because you're *never* going to have all the information you need to move forward. When it comes to making important decisions about your life or work, you need clarity, not certainty. And we had clarity then: we knew we wanted to own a company like Grand Circle Travel. People say, "Don't sweat the small stuff." We believe that. But we believe this even more: *never compromise on the dream stuff.* The dream is what makes life worthwhile.

Wager Big When It Matters

Harriet: In that split-second moment of decision, Alan wagered big. He offered Colonial Penn $9 million for Grand Circle, several million more than the company was worth on paper. Why? Because we saw the company's *potential*: it had a recognized brand, global reach, and loyal travelers in the retired American market. Over the years, we had become acquainted with many of the people who worked at Grand Circle, and we knew they were capable and experienced. We were confident that we could move the company forward quickly and profitably.

We also saw what *wasn't* on paper. Every mergers-and-acquisitions team knows that the value of a company includes more than its assets and bottom line. Value also lies in the opportunity the company offers to the buyer, the opportunity to achieve his or her own goals. All Saga wanted was Grand Circle's list of customers, but we wanted to make Grand Circle the leading travel company for retired Americans. Our dream was bigger, and this made Grand Circle more valuable to us than to Saga. Nine million dollars was the down payment on our dream, and we were ready to foot the bill.

Of course, Colonial Penn wondered if it could trust us to come up with $9 million. We were strangers to the company, and we were solo venturers; in contrast, Saga was a major European company with an official corporate presence in Boston. In other words, Saga looked like a safe bet, whereas Alan looked like . . . God knows what. The Colonial Penn people were dressed in Armani suits and silk ties, and across the table from them was Alan, a cheeky, sunburned guy in an ill-pressed suit. Our unsolicited offer made them nervous.

Alan sensed Colonial Penn's reluctance, but instead of backing down, he upped the ante. He wrote a $1 million check as a gesture of good faith, and he pushed it across the table. The offer was too good for Colonial Penn to pass up. Grand Circle was ours. Alan had the experience and the confidence to recognize that we could build this company into something great, and so he wasn't afraid to put his money where his mouth was.

If you truly believe in yourself and your own dream, you must be ready to make a grand gesture to demonstrate your seriousness. This is true whether you're trying to get a bank loan or funding from private investors, or trying to convince a distributor to carry your products, or working to persuade potential clients that they need the service your company is offering. Have the courage of your convictions, and you will begin making your own opportunities.

We were thrilled when Colonial Penn decided to sell Grand Circle to us. Unfortunately, Alan's former partner at Trans National Travel wasn't as excited, and he invoked the noncompete agreement. We could have fought him in court, but we had bigger fish to fry. We needed to focus on getting Grand Circle profitable, and so we negotiated a settlement with TNT, wrote another big check, and closed the door on the old days. Never allow yourself to be sidetracked by petty problems when you're starting a business— even if they're legal problems and even if they prove to be expensive. Keep your sights on growing your business.

Hit the Ground Running

Alan: We signed the purchase agreement with Colonial Penn on March 31, 1985. Almost before the ink was dry on the contract, terrorists fired a rocket at a Jordanian airliner in Athens. Two months later, a TWA flight was hijacked en route to Rome and an Air India flight was blown up in Irish airspace, killing 329 people. In October, the cruise ship *Achille Lauro* was hijacked in the Mediterranean and an American passenger was murdered on deck. In December, Arab terrorists attacked airports in Rome and Vienna. It was a dismal nine months for the travel industry, and Grand Circle was off to a very rocky start. We could have walked away from the deal with Colonial Penn early on, invoking the material damages clause of the purchase agreement. But we didn't. We believed we could move forward with Grand Circle and do well, and so we stayed the course—and negotiated a lower price for the company.

Few new ventures will face the kind of geopolitical threats that we did that first year, but every new business has to be ready to slay some dragons. When we counsel first-year start-ups, we often invoke what Harriet calls the "oxygen-mask" approach to business, reminding them that before they can thrive, they must survive— just as when you're flying, you need to put on your own oxygen mask and make sure you can breathe before you help the passengers sitting next to you. You need to determine what your most pressing issues are and deal with those immediately—or you'll be out of business before you even get started.

We believe there are only a few critical things that most new ventures need to do during their first year:

- Hire great people.

- Start making money.

- Focus on your company's best products or services.

- Find your core customers.

- Make tough decisions early on.

At Grand Circle, we had to tackle each one right away; so let's take a look at what we did during our first year of operation.

Hire Great People

Alan: Harriet and I needed help to get our new company off the ground. The two of us alone couldn't run a business that already had 120 employees and 16,000 customers, especially not with two toddlers at home, and so we set out to hire some key people.

Fortunately, during our early years in the travel business, we had come to know some of the best talent in the travel industry. Soon after we acquired Grand Circle, we gathered together an exceptional board of advisors and engaged some leading industry consultants.

I also brought in people from Trans National Travel, though they cost us a hefty price because we had to pay TNT to waive their individual noncompete agreements. One of those hires was Charlie Ritter, an old friend who'd been around the travel business for a while; he set up a crucial piece of technological infrastructure—our innovative computer system, fondly called GERT (Grand Circle Enters Revolutionary Technology). Other TNT defectors were Bruce Washburn, our first president, and Bruce Epstein, our new head of marketing. Jan Hobbs-Bailey came from another Boston-based travel company to manage operations, and Mark Frevert joined the marketing team. I loved these guys. They were smart, funny, competitive, crazy-hard workers with years of experience and lots of big ideas. Mavericks all, they threw in their lot with us, attracted by the opportunity to take leadership roles every day.

It cost us some money to get the people we wanted, but it was money well spent. We have always believed that great people are the key to a great company, and you should always be willing to beg, borrow, or steal them. You may not agree. But however you staff your start-up company, be ready to pay for the privilege of surrounding yourself with talent, energy, and optimism. Practice frugality somewhere else. Good people are valuable beyond measure.

Start Making Money

Harriet: We wanted to build a different kind of company, but first we needed to stop the red ink. When we bought Grand Circle, it had sales of $27 million a year, but it was also losing $2 million a year, more than $5,000 a day. Alan and I had bought the company with our life's savings (there was no outside financing), so that $5,000 a day was $5,000 of our own money, and our children's inheritance. It didn't take a degree in finance to draw the conclusion that we had to work fast before the money ran out or we wouldn't have a company to run!

Alan was ruthless about finding ways to stem the bleeding. He dismissed 60 Colonial Penn employees. He closed one office and opened another in a less expensive (seedy) part of town. He furnished the headquarters building with cast-off furniture. Most significantly, he took a hard look at Grand Circle's trip roster, with an eye to eliminating the low-performing trips.

Over the course of 12 months, we completely reinvented Grand Circle Travel. It was a wild ride, and we made mistakes. For instance, Alan closed the company's office in Long Beach, California, as a cost-cutting measure. Unfortunately, the associates who worked there were the ones with knowledge of the Far East and the American market west of the Rockies—knowledge we would need later, when the European theater was rattled by terrorist attacks and growing unrest in the Middle East.

We couldn't have predicted those developments, of course, but had we moved more slowly, we would not have faced that predicament. But moving slowly isn't something we do at Grand Circle. In fact, people often joke that we are "addicted to speed." The compulsion to keep moving is partly a reflection of our personalities. Alan is restless and hard-charging by nature, and I like to see practical results right away. It is also a reflection of the volatile nature of the travel industry, which changes every day. But I also think our obsession with speed, like our famously strong work ethic, might be the result of that crazy year when we were losing $5,000 a day.

Speed may not be important to every new venture. In your industry, you may be able to take the long view and slow course. But if you are an entrepreneur and you have money in the game, you must ask yourself whether you are you willing—and able—to work fast enough and long enough to keep it.

Focus on Your Company's Best Products or Services

Alan: No business can be all things to all people; you need to focus on what you do best, identifying the products that deliver the greatest value to your customers and the highest return to your stakeholders. In the travel business, the "products" are trips. When we bought Grand Circle, we inherited 500 different trips, which was exciting to us travel buffs until we discovered that more than half of them either lost the company money or were not profitable enough to keep us afloat. I told the product development team to get rid of them, and that caused some tears, because travel developers get attached to their trips. But once we eliminated those bottom-performing trips, our profit margin improved greatly, along with our personal bank balance.

If you have a new venture, you will need to identify your company's best products or services, no matter what business you're in. Focus on whatever is most profitable and provides the most value to your best customers. If you stay close to your company's Extreme Competitive Advantages (which we'll say more about later in this chapter), your business will not only survive, but thrive.

Find Your Core Customers

Harriet: Once you identify your business's best products, you need to figure out who your best customers are—or should be. For us, it was easy. We had bought Grand Circle from AARP, an association dedicated to the needs of retired people. We continued marketing directly to the list of former AARP travelers, eventually identifying our prime market as American travelers over the age of 50—not mere tourists interested only in seeing the usual sights, but serious travelers who really wanted to get to know other parts of the world.

Now that we had identified our core customers, we needed to figure out what made them happy. What did they like and dislike most about our trips? Grand Circle had been doing post-trip evaluations for years, even before Alan and I bought the company, and its travelers were famous for giving candid feedback. What a godsend—we inherited a built-in product quality system! We read those evaluations closely, and the trip developers changed the trips as soon as we saw patterns in the surveys. For example, we stopped hustling our customers from place to place, and we slowed down the trips so the travelers could better appreciate the people and cultures in our destinations. We also gave our customers more value for their money by cutting prices, eliminating intermediaries overseas, and adding new features to the best trips.

Customers are the reality check that every new venture needs. Too many entrepreneurs make the mistake of believing that people will want whatever they're selling. But that's not how the market works. The first rule of business is to find an unmet market need—and then fill it in outstanding ways. Whatever your business is, you need to find out what your customers want most from you, and then you need to make sure you deliver *that*.

Make Tough Decisions Early On

Alan: I believe it's best to make major decisions about how to run your company early on, before those decisions get even harder to make. In our case, the hot issue was where to headquarter the company. When we bought Grand Circle, it was based in New York City. But Harriet and I lived in Boston. That was where we had both grown up and where we settled when we married and started our family.

The usual advice to a takeover team is to limit the number of changes in the company's takeover year. *Stay put. Go slow. Let the dust settle. Give yourself time to figure out what works and what doesn't before going off half-cocked.* I know this is the rule because we broke it repeatedly after buying Grand Circle, and I heard the tongues clacking each time.

Unfortunately, after only six months, I knew all too well what wasn't working for us, and it was the location of our business—and the commuting. Every Monday morning, I would catch the early-morning commuter flight to New York with the rest of the Boston-based leadership team, and we wouldn't return until the end of the week. Harriet stayed at home in Boston, fuming. This wasn't getting us closer to our dream. We had two small children who deserved the attention of both parents. Besides, we wanted to be together, to work as a couple. Ours was a shared dream, and we knew that our strengths and weaknesses balanced each other's. The decision was obvious: either we had to move the family to New York City, or we had to move the company to Boston.

Every advisor—business, legal, and personal—told us to leave the company in New York for at least a year. But we had grown up in and around Boston, and our friends and family were there. We also knew we'd be lost in New York. And we hadn't gotten this far by doing things the way people told us to do them. We reminded ourselves: *never compromise on the dream stuff*. And so the decision was made; Grand Circle would move to Boston.

Shortly before our first anniversary, we moved Grand Circle Travel from a skyscraper on Madison Avenue to a crumbling warehouse in a deserted section of South Boston. It is a lively and picturesque district today, but at the time, the neighborhood was separated from downtown by a polluted canal and piles of construction rubble. Nevertheless, our new home at 347 Congress Street had advantages. It was big, it was cheap, and it had a real Boston feel. Easy access to South Station made it convenient for commuting associates (the commute wasn't over, of course; New Yorkers would now be commuting to Boston). Street parking was also plentiful then, and the office was close by our home. Now we could both work on building a great company and be home for our kids when they got home from school.

Our new South Boston office was a great fit for the kind of company we were building. We wanted to promote open minds and open communication, to create a space where people saw themselves as leaders and contributors. Traditional corporate architecture—big offices, fancy conference rooms, closed doors, and cubicles—is designed to do the opposite, to keep people in their place. Our six-story building was nothing like that. Furnished with used furniture and covered in century-old brick dust, it wouldn't let us take ourselves too seriously. Our open plan encouraged people to speak up and feel safe in challenging others, including us. Our casual new digs in this funky part of town helped us create a freewheeling culture committed to doing something great.

Our move wasn't entirely smooth, though. We made a big mistake along the way. When we decided to move the office from New York to Boston, I made a list of 28 top-performing people in New York, with the idea of offering each of them a great package if they would transfer to Boston. We knew that New Yorkers are not always fond of Boston, and we were worried about how they would take the news. So we delayed telling them. It didn't occur to us that New Yorkers might read the *Boston Globe*, which reported our intentions one Sunday morning.

When we arrived in New York that Monday, we were greeted by associates wearing pig snouts. This was our first big mistake, but it taught us a crucial lesson. If we wanted a different kind of company, we had to be forthright and honest with our associates. It's sometimes painful, and we've stumbled occasionally, but since that day, we've delivered bad news as quickly and directly as possible. We learned our lesson the hard way, and we're passing this advice on to you so that you won't have to. Be open and truthful with your employees about what's going on in your company from the very first day. In doing so, you will develop trust and loyalty among your employees, who will, in turn, make your customers happier and your business better.

No matter what business you're in, you're going to face some tough decisions right out of the gate. You might have to part ways with one of your founding partners. You might want to go in a different direction from the one you and your team originally envisioned. You might realize there's no market for a product you've spent a ton of money developing, and you have to abandon it before you start throwing good money after bad. Whatever the decision, don't delay. Starting a business is hard enough. Make the tough calls early, and you will save yourself trouble in the long run.

■ ■ ■

That first year was fun, wasn't it, Harriet?
That first year was hell, Alan, and you know it.
Yeah, but a fun kind of hell.
Well, maybe it was an adventure.
OK, an adventure, then. "Our Grand Adventure."

■ ■ ■

Extreme Competitive Advantages

Alan: Right from the get-go 26 years ago, we knew we would need to build an organization that was very different from traditional "corporate American business." Our choices of unconventional and open office space in a funky Boston neighborhood were a good first start, but to be successful, we knew our company would need to be more than just "a cool place to work." If we were going to succeed in the chaotic world of international travel, our company had to build a unique organization that could thrive in crises. So we set about developing our Extreme Competitive Advantages—our strengths, the things that would set our business apart from everyone else.

We believe that, to be successful, a business (*any* and *every* business) needs to focus on what it does best. We've learned (the hard way) that you can't be all things to all people—but you can waste a lot of time, money, and energy trying. We believe it's important to put your efforts into areas where you can be world class so that you have a competitive edge. We know it is tempting to do whatever work might come your way, but our experience says, "resist!"

We've seen this happen to so many other businesses. We've seen it happen to a friend who owned an architecture firm that specialized in designing hospitals. But during a slow period, he took on a client who wanted to build a new office building and a wealthy couple who wanted to build a Tudor-style home. Then he lost out on the next major hospital because his firm was too busy with the smaller clients it took on when business was slow. We've seen it happen to an auto mechanic we know who's an expert in vintage car restoration but who takes on customers that need minor repairs to their modern cars; then he doesn't have the time or resources to take on much more lucrative (and fun) restoration projects when they come his way.

We've seen this happen to our business, too, when we first bought Grand Circle. When we acquired it, Grand Circle had 500 trips of all sizes and for all types of travelers. That's a lot of products for a small business, and "marketing to everyone" is certainly *not* an Extreme Competitive Advantage! We knew the first thing we needed to do was focus on what we did best, which was to offer international vacations for Americans over age 50. (Of course, we didn't figure that out immediately, and we'll tell the whole story of how we developed our vision, mission, and values in Chapter 2.)

Over the past 26 years, we've developed six areas of dominance—*our* Extreme Competitive Advantages—that are so strong and so far advanced that we know it would take our competitors at least three years to replicate them—*if* we stood still . . . and we never stand still. We constantly work to hone our Extreme Competitive Advantages, in good times and in bad, because we need to continue to do what we do well in order to remain ahead of the competition.

GRAND CIRCLE'S EXTREME COMPETITIVE ADVANTAGES

1. Associates are number one
2. Integrated corporate philanthropy
3. Unsurpassed value
4. Measure for excellence
5. Focus on the lifetime value of customers
6. Focus on niche market opportunities

The rest of this chapter gives a quick overview of each of our Extreme Competitive Advantages—and the rest of the book describes them in even more detail. We hope these descriptions give you ideas about what *your* business does best, so that you can stay focused on that, on what gives you a competitive edge, as you're growing.

1. Our Associates Are Number One

Harriet: Many businesses say "our customers come first." At Grand Circle, our associates do. Our associates are number one, and we call them *associates* instead of *employees* because we don't subscribe to the traditional employer-employee top-down relationship. We are a team of people working together. We believe—and it's proved to be true for our business—that if you value the people you work with and you treat them well, they will value your customers and treat *them* well. You take care of your associates, and they'll take care of your customers.

Also, if you empower your associates, you'll develop leaders who can handle almost any situation at any time—and in the international travel industry (and many others), that's an increasingly important skill to have. Many businesses don't have the luxury of having only their "top people" make decisions to handle problems as they arise . . . you need *all* your people to be *top* people.

Leadership development for all our associates is critical to our success as an organization. We've devoted Chapter 3 to how we develop and sustain a culture of leadership from anywhere.

2. Integrated Corporate Philanthropy

Harriet: We also believe corporate philanthropy should be an integral part of every company's business, not something you do *after* you've become successful. We believe our commitment to strategic philanthropy is one of our competitive advantages—and one of the key drivers of our profitability. In the early days of our company, we donated money and volunteered privately; then we began encouraging our associates to volunteer and provided information on organizations that we knew needed help.

Later on, in 1992, we created the Grand Circle Foundation to formalize our philanthropic work around the world and since then have donated or pledged more than $50 million to educational, humanitarian, and cultural causes. I can tell you that we have a better and more profitable company, with more committed associates and more loyal customers, because of the work we do through our foundation. Chapter 4 describes what we do and what we learned not to do, and offers advice for others who want to give back through their businesses.

3. Unsurpassed Value

Alan: Our customers know us as the company that delivers wonderful trips at a great price. We call this Extreme Competitive Advantage "unsurpassed value." No one can offer travelers the tremendous value we do in our vacations.

Over the years, we have found many ways to keep our prices down. We buy direct wherever we can (and we usually can), thus bypassing local ground operators. We market directly to our customers, eliminating travel agents and their commissions. Instead of prospecting for new customers, which is expensive, we cultivate our repeat customers and their referral business. We own and operate our own ships; we keep our load factors high; we negotiate favorable contracts with vendors who want our high-volume, year-round business; and we scoop up inventory when our competitors dump it in a crisis-induced panic. Chapter 5 provides details on how we deliver unsurpassed value to our customers.

4. Measure for Excellence

Alan: We also strive for excellence in every one of our trips, and we measure our performance against it. When some prospective investors jokingly called us "maniacs on excellence," we sent them packing, because they just didn't get it. Mere trips are a dime a dozen, but *great experiences* take constant tending. More important, great trips ensure repeat business—which is the engine driving our business, as you'll see below in point 5 and later on in Chapter 6.

To help us measure for excellence, we send our customers a 12-page survey with more than 90 questions. The surveys are waiting for them in their mailboxes when they get home from the trips, because we hope they'll answer them while their experiences are still fresh in their minds. The categories of questions are the same on each survey (based on the brand—that is, GCT surveys are slightly different from OAT surveys), but each survey is specific to the traveler's particular trip in terms of rating specific hotels, Grand Circle Foundation sites, and other particulars.

We receive about 200 post-trip surveys *every day*. "Quality" is one of our six core values (we'll discuss values in Chapter 2), and we track it assiduously: chasing down lapses, following every lead for improvement, and canceling entire programs if our customers tell us we are on the wrong track.

If you want evidence to back up our claim that our associates are taking good care of our customers, just take a look at our customer surveys. In 2010, we had 115,000 customers, and 70 percent of them filled out our very detailed surveys. That 70 percent translates into about 80,000 people. How many of those customers rated us "excellent"? Here are the results:

- 83 percent of Overseas Adventure Travel customers

- 79 percent of Overseas Adventure Travel Small Ship customers

- 80 percent of Grand Circle Travel customers

- 80 percent of Grand Circle Cruise Line customers

- 81 percent of our customers overall

We rely on these surveys to make sure we're keeping our customers happy. Many business owners are unwilling to invest in evaluating their products; instead, they gauge their products' success solely on sales figures. But we've learned that by surveying our customers, asking them very specific questions about very specific aspects of our products, we're able to garner invaluable information about what our customers like and dislike—and how we can improve. Surveying your customers—whether in focus groups, by e-mail, via mailings, or by whatever method works best for you and your product—will pinpoint where you're pleasing your customers and where you're not. Their input is, as the MasterCard ad says, *priceless.*

5. Focus on the Lifetime Value of Customers

Alan: Most of our customers love us, and they travel with us again and again. On a typical trip, more than half the customers have traveled with us before. Many take several trips a year with us, and 33,000 households have traveled with us more than six times. So many of our customers love their experiences with us so much that they refer between 30,000 and 40,000 new customers to us every year. We have a mutual admiration society of the very best kind, and we honor it by paying very close attention to the customers we already have and by serving them well and fostering loyalty.

We discuss customer loyalty in detail in Chapter 7, but here's a quick overview. Most other travel companies take the "cast-a-wide-net" approach, always looking for new customers in the widening pool of retiring baby boomers. We traveled down that road for a long time ourselves. But we've found that our more personal approach, although it's more time- and labor-intensive and more costly, is more profitable—and more satisfying. We feel we are building a company of active, like-minded travelers, people who believe (as we do) that travel is a joy and a privilege. We believe we have an opportunity not only to grow a business but also to make people's dreams come true—and maybe even help change their lives.

We also know who our core market and customers are: two-thirds of our travelers are women. We've been doing demographic and other market research since Mark Frevert joined us right after we bought the company, which we describe in Chapter 8. Even though we have a database of more than 4 million names, which we can segment by household, we really learn what our customers want by *talking to them personally*. And by *listening*. Hardly a radical idea, but we believe so few businesses do this.

In our early days, we took as many of our own tours as we could, and we talked to as many customers as humanly possible. And we still do. In the past few years, Harriet and our daughter, Charlotte, who is our vice president of interactive marketing, have met with hundreds of women to ask about our business practices, their travel preferences, and improvements we could make. We sent surveys to many of these women before we met with them, but that face-to-face discussion proved to be the most valuable. At one meeting, two years ago, we had to limit the number of women attending to 225—and another 50 women showed up anyway! It was a great day of shared discussion, questions, recommendations, laughter—and very direct feedback.

Our company is fortunate because we can communicate directly with our customers since they "buy" directly from us. However, even if you sell your product or service through an intermediary—if you use a retailer or if you serve as a subcontractor to your ultimate customer—you can still find a way to communicate directly with your customers. If you sell your product via retailers, you might include a prepaid comment card in your packaging, perhaps offering a discount off the next purchase for filling out the card. If people register your product, you could have an online survey.

Whatever your business is, it's so much easier to work with repeat customers who love what you do than to continually search for and try to please *new* customers. So from the very start, you should strive to develop customer loyalty. Don't take your customers for granted—or they won't be your customers anymore.

6. Focus on Niche Market Opportunities

Alan: The international travel business is extremely competitive, and world events compel the industry and those of us working in it to constantly change and react. We have always focused our business on identifying the niche areas where we can excel, where we'd have a competitive advantage, versus competing in areas where we're pretty sure we cannot excel. Over the years, we've developed several marketing strategies that have set us apart from others in the industry. One is to focus on small-group travel through our travel brand Overseas Adventure Travel; another is to operate only small-ship (versus larger) cruise products and to deliver them better than anyone else at unbeatable prices. We'll provide more detail in Chapter 8, but in our company, small has helped us get big.

In terms of small-group travel, Overseas Adventure Travel, which we acquired in 1993, is the undisputed leader—on land and on water. This is an enviable place to be because small-group travel is the fastest-growing segment of the travel industry. We are adventure travelers ourselves, and we understand the allure of out-of-the-way places, intimate perspectives, and close camaraderie on a trip. We also know that a successful small-group trip is not a big-group trip cut down to size; it is a whole other experience. We have spent 17 years developing itineraries that really take advantage of small-group size (no more than 16 travelers on OAT land trips, and no more than 25 on OAT ships), and OAT's small ships are now ranked among the best in the world.

We believe small-group travel is the wave of the future. As the world gets smaller, travelers will want to take a closer look at places that are off the beaten path, where they can engage personally with fellow travelers and with local people going about their daily lives. Our curious, convivial, independent-minded OAT travelers are already traveling that road—more than 50,000 of them a year.

In addition, we now specialize in small-ship cruises—something we never would have predicted when we first started out in this business! Harriet has never been a big fan of boats, and my prior experience with water was chiefly from the seat of a lifeguard's chair. But today, Grand Circle is a leader in the small-ship cruise market. We didn't quite know what we were getting into when we bought—and then built—our first ships in the late 1990s. But we learned fast, and we now own and operate a fleet of 60 small ships all over the world.

Our advantage comes from two hard-won achievements: complete control over the ships' operations—outfitting, staffing, provisioning, and service—and the integration of our signature land tours. It's a win-win for our customers—and a bargain, too, because we can price our cruises $1,000 to $2,000 less than our competitors do.

Small-group travel and small-ship cruises are just two of our niche strategies that have helped our company succeed and grow. Obviously, they're specific to the travel industry, and so unless your company is competing with ours, these market strategies won't work for you. But whatever business or industry you're in, we believe it's critical to identify what *your* niche market strategies will be that will set you apart from *your* competition.

While we use our Extreme Competitive Advantages to succeed in a highly competitive industry, we are constantly required to react and respond to crises. We operate in one of the most unpredictable industries of the twenty-first century—the international travel industry, where problems and crises are the norm. You can't just create a few itineraries, pick some hotels, book some passengers, and then sit back and relax. In the travel business, *something always goes wrong*. Airline workers go on strike; travelers get sick; hotel reservations mysteriously disappear. Beyond these difficulties, real calamities are always happening in some part of the world—as we saw during our first nine months in business. Even 25 years later, we still see that, and so we've devoted Chapter 10 to how we manage crises.

To give you a brief glimpse of what we deal with every day, consider just one recent crisis we faced, in 2010. You wouldn't think a volcano in Iceland would disrupt international travel, but boy, did it ever! When it erupted, it spewed 330 million cubic tons of ash into the atmosphere. That cloud shut down all airports in 20 European countries. Those closures grounded 100,000 flights and stranded 10 million passengers for six days. They also stranded anyone who had a connecting flight to Europe that originated in the United States, the Middle East, or Asia. And all of that affected thousands of our customers, of course. What do you do in a situation like that? Read Chapter 10, and you'll find out what we did and how our customers reacted.

Of course, there will always be unexpected world events that impact myriad industries. The world is an increasingly unpredictable place. But as the Viking saying goes, "What doesn't kill you outright makes you stronger," and we believe that how we handled this challenging year did indeed make us stronger. Most organizations can survive in predictable times, but many struggle to stay afloat during times of volatility. At Grand Circle, we've developed the ability to not only survive but thrive in crises. In fact, we made that ability into one of our core competencies. We believe our company is a great model for anyone who is trying to build an organization in the twenty-first century.

Why Our Competitors Don't Really Worry Us

Alan: Finally, we don't worry about competitors sneaking a peek at this book and trying to copy our Extreme Competitive Advantages. In fact, they try to copy them all the time. We don't worry because we know how long it took us to consolidate these advantages, and we know it would be very difficult to implement them outside of our unique corporate culture. If our competitors make a concerted effort, they might be able to replicate what we've done in, say, five years. But by then, we'll be way down the road.

Now, how about you? What are your Extreme Competitive Advantages? What do you do better than anyone else, in terms of your market, your customers, and your product line or services and with respect to your competition? If you focus only on your competitive strengths, you are on your way to building your successful company.

To learn more, please read on!

Achieving Success by Adhering to Mission, Vision, and Values

Define your vision, mission, and corporate culture early on—these will serve as the framework upon which to build your dream.

Harriet: We're well known in the business world for our corporate culture. Not just for our "Grand Circle style," which is kind of wacky, fun, and irreverent, but also for the way we pay attention to our vision, mission, and company values.

In the mid-1980s, when we bought Grand Circle, corporate America was experiencing a craze of visioning exercises, mission statements, social responsibility, and values-driven management. Most of the business leaders we knew dismissed it all as a bunch of idealistic, time-wasting bull***. Even President George H. W. Bush made light of "the vision thing." But we thought the vision thing was great. Alan and I were on a mission to help change people's lives, and we were proud of it; so we put a lot of time and effort into building a culture that would get us there.

We wrote mission statements. We argued about accountability. We took several Outward Bound trips. We formed a department called People, Culture & Corporate Responsibility. Alan and I even went on a modern-day vision quest, spending three days and two nights in total silence together on a vast plateau in Arizona— just to get our priorities straight. Go ahead, call us crazy. Call us the "culture club." Many of our associates have also been skeptical, but never for long, because our strong corporate culture makes its benefits known right away.

We believe a carefully considered culture can be the difference between business success and failure. This chapter describes how we took Grand Circle to the next level of success by clarifying the company's vision, mission, and values, and we offer suggestions for how you can do this in your business, too.

Culture Will Drive Your Enterprise, Whether You Like It or Not

Alan: I do a lot of leadership consulting work, both for our business training company, Grand Circle Leadership, and at Babson College, which houses our Lewis Institute for Social Innovation and Entrepreneurship. I meet a lot of smart, ambitious, high-powered people in this work, both business leaders and social entrepreneurs. Really brilliant, hard-working people. So it always surprises me how often I hear them say things like, "Oh, we don't have an actual mission statement," or "We're not into vision and values. We're just here to work." As if corporate culture were somehow *optional.*

Culture isn't optional. Not every organization talks about its culture, but every organization has one: every business, every nonprofit organization—every family, for that matter. You've got to name it, own it, and manage the hell out of it. If you don't, then one of two things is going to happen: either the strongest personality in your organization will take over the company, or the unspoken culture will start running the show. That's not always a bad thing, but the potential for harm is very high, especially in organizations that are strongly driven by profits, sales targets, quotas, bonuses, and other money matters. In those organizations—and that includes most companies—the underlying culture can get pretty nasty.

I've seen a lot of greed, secrecy, and sabotage get loose when leaders fail to tend to their culture. We've seen similar behaviors at Grand Circle when Harriet and I took our focus off our corporate culture. Better to be proactive, to decide the vision, mission, and values that you want to guide your company before you find yourself on the losing side of a culture war.

Your Company's Culture Affects Everyone

Harriet: One of the reasons we've always made such a big deal about culture at Grand Circle is because we're a travel company. Our business is selling the experience of different places and different cultures, and so we know how powerful—and divisive—culture can be. It is a huge, driving force that connects people and shapes their beliefs, behaviors, and way of life. Culture does that for every human group—Christians and Hindus, Hutus and Hottentots, Boy Scouts and Hell's Angels. It's the key to people's hopes and dreams, and it has a huge influence on their behavior.

OK, you probably see where I'm going with this.

Even before we bought Grand Circle, we knew we wanted to build a different kind of company, a company where our associates would look forward to coming to work, where they could grow into leadership roles, and where their work with us would enrich their personal lives. We also wanted to create trips so amazing that they would leave a lasting impression on our travelers' lives. That's what Alan and I talked about on the beach on Captiva Island. Those were *our* hopes and dreams. Clearly, we needed to build a culture to make that happen, because—as every traveler knows—it's in the culture where the magic happens.

Maybe your enterprise isn't as strongly driven by a dream or a social mission as ours is. Maybe you are more driven by the bottom line. You still need to pay attention to culture, because while it isn't fashionable to say so in the new global village, some cultures are *stronger and more successful than others.* What your company does about culture matters. It can either lift your boat or sink it.

You Set the Tone

Alan: We understand a lot about culture now, but it took a long time to really get it. Our first year was so crazy, it was hard to focus on things like vision and mission and values. To a great extent, Grand Circle was being driven by a group of energetic young people who didn't quite grasp that what they were doing was difficult. Those were fun times. The staff was creative, rambunctious, and irreverent. No one paid attention to protocol or etiquette. We laughed a lot and played games, competed openly with each other, and ignored rank. It was wonderful: youthful, energetic, and bold.

This early culture had a lot to do with Harriet and me. We were young ourselves, in our mid-thirties; we dressed casually and had a freewheeling, traveler's lifestyle. We loved challenges and hated sycophants, and both of us loved to compete. This was exactly the "different kind of company" we had hoped to build. We wanted to attract independent-minded, high-performing people, put them to work in an open, demanding, and fun environment, and watch them grow into leaders. As it turned out back then, we succeeded beyond our wildest dreams.

Our biggest fear at the time was that we might lose our maverick style. We worried that as we prospered and grew, we'd become structured, complacent, and ordinary. We'd become just like everyone else. While some young companies look forward to joining the staid ranks of corporate America, we believed maintaining our irreverent, hard-charging culture was the key to building a truly great organization.

But how could we channel all this energy and independence in a more productive way? How could we make our different kind of company extraordinary and still keep our maverick spirit alive? I knew we had to work on vision and mission, but where to begin?

In 1986, shortly after we moved Grand Circle to Boston, we issued our first Boston-generated brochure. It was titled "Faraway Places," and it had a picture of the Taj Mahal on the cover, along with the tagline "Leadership in travel for active mature Americans." Unlike other travel brochures of that era, the photograph didn't show smiling tourists; instead, it showed pilgrims approaching the temple.

The cover of that brochure summed up our entire business model at that moment in time. In a kind of distilled simplicity, it defined both our product and our target market: our customers would be Americans, we would design our trips for active seniors, we would go to faraway places, and we would provide opportunities for our travelers to engage with the people and cultures at our destinations. The brochure cover was, in effect, our first vision statement.

Around the same time, Mark Frevert started tacking up handwritten posters around the building, urging associates on with the words "Fast, Flexible, and Time-Competitive!" The slogan didn't exactly roll off the tongue, but it summed up our expectations for job performance. It was our first go at a values statement.

Of course, we would need more than a brochure and a slogan to give direction to our fledgling company. We would need to make our product strategy, vision, and values more explicit. But what strikes me now is how close those early formulations were to the formal statements we made later.

It's hard to tackle corporate culture when you're scrambling to keep all the balls in the air, and it's easy to make mistakes. Some companies arrive at ridiculous mission statements by accident, treating some offhand remarks by their founder as commandments. Others print up some flowery words in curlicue script, hoping they will somehow inspire folks. They hardly ever do. You can't import a culture from outside the company; it has to come from the guts of the enterprise.

We made lots of mistakes ourselves. One of our first values was "thriving in chaos," but at times, we let chaos reign and derail the business. We allowed associates to create and then test a wide range of new trips by land and sea—pretty much whatever they felt might work in an ad hoc, disorganized, and chaotic manner—and as a result, we lost about $500,000 over a two-year span on poorly executed trips. Over time, as we grew, we abandoned "thriving in chaos" for "thriving in change," and this value has stuck.

So go ahead: start small. Think about the slogans, inspirational remarks, company jokes, and customer tributes that resonate with you. Go back to the heroes of the company, and ask some trusted associates what really fulfills them about working for your organization. But most of all, remember why you got into the business in the first place. Write all these things down, and then just wait—because pretty soon some crisis will come along that will turn those jottings into a vision, mission, and values to live by.

Defining Your Business

Alan: Our first vision crisis came in June 1988, when I took the top leaders of Grand Circle on a six-day Outward Bound–facilitated trip down the Salmon River in Idaho. We did a lot of wilderness retreats in those days, but this one promised to be especially exciting because the river had a fantastic nickname: "The River of No Return." We thought that was great. We were young and cocky, eager to challenge ourselves. We intended to have a good time, but the business goal was to work on product strategy and to think harder about our overall vision for Grand Circle.

The white-water rafting was fun. Defining a vision for our future was not. Almost immediately, we got in a wrangle about the kind of trips we wanted to sell, and the group quickly split into two warring factions. One group wanted to sell high-volume, low-cost sightseeing trips that covered a lot of territory in a short time. If you've seen the movie *If It's Tuesday, This Must Be Belgium,* you know the kind of drive-by, mass-market trip they had in mind. (And if you haven't seen the movie, it's about a bus trip that takes travelers to nine countries in 18 days—not a great way to appreciate Europe!) The other group wanted nothing to do with whirlwind tours; they wanted to offer authentic cultural experiences at great value, trips that might actually change people's lives.

The discussion got ugly. Someone accused the mass-market people of peddling "cheap and nasty holidays." They were, in fact, very good at selling that kind of trip, and they had contributed to our dramatic company turnaround. But that wasn't the kind of travel experience I wanted to offer at all. It was nothing like the up-close-and-personal trips that Harriet and I had talked about on the beach on Captiva Island. This conflict was the 200-pound gorilla that had followed us from Boston to the Salmon River, and before long it was snarling.

To make matters worse, the rafting got scary. The river-safety expert who guided the group was thrown from the raft into the rapids for the first time in his 12-year career. One of our colleagues was also tossed out of the raft—and sucked into a whirlpool. Finally, during an intense evening debriefing, my old friend Charlie Ritter picked up a burning log from the campfire and began threatening our facilitator, Bob Gordon, a leadership expert from Outward Bound.

The ambitious agenda and the raging river had created a perfect storm. Suddenly, all the tough issues we'd been tiptoeing around for months came torpedoing up to the surface. I could see we had a major misalignment of goals, and I wasn't about to back away from it, because disagreement about goals can lead to serious business trouble. I pushed harder for a vision statement to guide our product strategy, and by the end of the trip, we had one. Rejecting "cheap and nasty" vacations, the group committed to a high-value, high-excellence model that could provide authentic travel experiences at an affordable price without sacrificing quality. We also committed to the idea that Grand Circle would strive to become a great company, one that would help change people's lives.

Unfortunately, the entire group wasn't truly committed, and shortly after we returned to Boston, both the company president and the vice president for product development left the company. Like many travel executives of that era, they believed our vision was too idealistic and that our product strategy would fail. We think time has proved them wrong.

Your Vision Should Inspire People to Greatness!

Harriet: The consensus we reached on the Salmon River became a vision statement that committed to excellence, profitability, value, cultural engagement, respect for our customers, and fulfillment for our associates. The vision has gone through several iterations since 1988 to reflect changes in our customers' travel tastes (much more sophisticated and adventurous) and to embrace important new ventures (adventure travel, small-ship cruises, and philanthropy). But while our strategies, tactics, and trips have changed, our vision has remained steady. Here's what our vision statement looks like today:

GRAND CIRCLE'S VISION TODAY

> We will strive to be the world leader in international travel, adventure, and discovery for American travelers over 50, providing impactful intercultural experiences that help change our customers' lives.

> We will establish this leadership position in the travel industry through application of our direct-marketing expertise and through the synergies gained from our acquisition of key nature and adventure travel companies.

> We will generate significant annual growth for our companies. Each year, up to 5 percent of our after-tax profits shall be given to fund our corporate philanthropy, Grand Circle Foundation, which supports the people and places our customers visit.

> We will strive to be a great company, providing a workplace that is stimulating, fulfilling, and meaningful to our associates, an environment where associates connect to their passions and to each other as we achieve our professional and personal goals.

We know our vision may be bigger than most. We have never hedged our dreams. In fact, part of our culture is daring to undertake seemingly impossible things, confident that the encounter will be exciting no matter what. Ours is a true traveler mentality: there are no boundaries, only new frontiers.

Trust in Your Vision—Don't Stray

Alan: A good vision statement is a compass that can actually lead your business and help you navigate. In our case, once we got rid of the "cheap and nasty" vacations, we were able to develop trips that satisfied the vision requirement that we help change people's lives. We did this by building them around what we call "unforgettable experiences."

Honey Streit-Reyes came up with the idea of unforgettable experiences in 1989, shortly after we got back from the Salmon River. We've known Honey since before we bought Grand Circle, and we hired her as our first European buyer. Honey also designed many of our early trips in Germany, Austria, and Switzerland. She was a genius at ground delivery, famous for fashioning hour-by-hour itineraries that were designed to wow the customer with interesting and unexpected encounters that they couldn't get with any other travel company.

Harriet: Honey designed our very first The Best of Eastern Europe trip in 1991. The Berlin Wall had come down just 18 months earlier, and so this was new territory for American travel companies. It was a really exciting time; anything seemed possible. Honey had grown up in East Prussia; she knew firsthand how to make the trip unforgettable—she booked our travelers into family-owned hotels, fed them home-hosted meals, took them to little schnitzel houses, and got them special deals in local markets. She stretched the trip out to 18 days so there would be enough time to really engage with local people and their ways of life. While other companies were herding their customers through empty cathedrals and tourist-trap *biergartens*, our travelers were visiting secret way-side shrines and talking to lifelong communists. It was like having a backstage pass to a rock concert.

Alan: We gave Honey a lot of freedom, not only because she was a trusted friend, but because we knew her work fulfilled our vision for the company. That's the true value of a good vision statement. It gives your company a compass and frees your employees to do their best work.

Next Up: Defining Your Mission (How You'll Do Business)

Alan: Vision and mission are different. At one point, maybe 20 years ago, we developed a set of complicated Venn diagrams to show how our vision, mission, and values were related. I think we related them to body, mind, and heart, as well. We were probably overthinking things. Today I just explain it this way: a vision statement says what a company hopes to become, a mission statement says how you'll do business, and values are the rules you play by.

Our mission statement came a little later than our vision, in the early 1990s. It was the result of some very deliberate work with our longtime friend and consultant Bob Weiler. Bob had been executive president at the Hurricane Island Outward Bound School, and like me, he believed in the power of mission statements to direct corporate action in meaningful ways.

Compared with other corporate missions, ours is a little unusual. It doesn't define a goal; instead, it outlines four responsibilities. Here it is:

GRAND CIRCLE'S MISSION TODAY

Grand Circle Corporation is committed to a mission that creates a balance between our responsibilities to our customers, our associates, our shareholders, and our world.

› *Associate Responsibility*
  We will provide an environment that fosters professional development and encourages personal growth for our associates. We will maintain competitive compensation and benefits packages relative to the industry and community. We will conduct business with respect for each individual and his or her role within the organization.

› *Customer Responsibility*
  Grand Circle Corporation, through direct marketing, is committed to providing active, mature Americans over 50 with the most exciting travel, adventure, and discovery programs in the world at unequaled value. We strive for 100% customer satisfaction.

› *Financial Responsibility*
  We will operate Grand Circle Corporation in a sound financial manner, to create growth and increase its value.

› *Social Responsibility*
  Global citizenship is central to the success of Grand Circle Corporation. We will commit time, people, and funding through Grand Circle Foundation to local, national, and global communities in which we live and explore, thus creating a better world for our travelers to discover.

It wasn't easy getting our mission statement down to 178 words. We argued about a lot of things, especially about which responsibility should come first. In the end, we made our associates number one, a decision that would have a profound effect on the company. Of course, your mission statement will be different from ours, because every company's mission is unique to that business. But having some sort of mission to which everyone in your company is committed is critical. Spend some time on it. Make your priorities clear. Your success will depend upon how you achieve what you say you're going to do.

You don't need to bring in an outside consultant to help you. What you do need to do is think about what your company's goals are—not quantitative goals like how much revenue you want to bring in or how quickly you want your company to grow, but qualitative goals like how you want to do business and what you want to achieve.

Who Cares About Values?

Alan: Sometimes, when I give a talk on values to people outside the company, I can see people tuning out. They roll their eyes or start fiddling with their BlackBerrys. Once a guy in the front row actually snuck an earbud into his ear and started listening to his iPod. I confronted him.

"Hey, buddy?" I asked. "Why aren't you listening to me?"

The guy was surprised, but he answered right back.

"This is where you start talking about peace, love, and understanding, right? How you can't be a greed head, gotta work together to help your fellow man? It's the 'Mom talk.' We get it a couple of times a year from the HR department."

I had to laugh. When did values get such a bad rap? But I felt sorry for the guy, too, because clearly his company didn't have a clue about leadership and management.

Your company values are not fluff. They are the rules you work by. They tell you how to approach every task you encounter—every market decision, every customer complaint, every crisis. When your values are disconnected from the work of the enterprise, you're sending your people out without a playbook.

The Six Values That Made Our Company Great

Alan: Grand Circle has six values, which we identified over a period of about five years. They don't much resemble the values any of us knew at our previous jobs. For example, Mark Frevert came to us from a Fortune 500 company where the chief values were accuracy and fiscal conservatism. At Trans National Travel, my first company, the top three values were profit, profit, and profit. It's one of the reasons I left. In comparison, Grand Circle's values seem revolutionary. Here they are:

GRAND CIRCLE'S VALUES

1. Open and courageous communication
2. Risk taking
3. Thriving in change
4. Quality
5. Speed
6. Teamwork

Your company values will be different, of course, and you can't pick them out of a catalog. They will depend on your organization's industry or sector, the demands of your market, and the way you structure your workforce. Your values might be profitability and growth. They might be honesty, hard work, and humor, or passion, performance, and profits. The important thing is that they be directly tied to your vision and mission and that your employees know exactly what they are.

It is not enough to identify your company's values; you must also specify how they will play out in the workplace. Associates can't be left to guess what open and courageous communication—or teamwork, or loyalty, or love—might look like. It is leadership's job to make that clear, in terms that are explicit, achievable, and measurable. Only by operationalizing the values can you make them part of your everyday business practice.

Let's take a look at Grand Circle's values, one by one.

1. Open and Courageous Communication

Alan: This is probably our signature value. It's definitely the one that outsiders most often remark on. We know that the combined intelligence of our organization is astounding, but it is valuable only if it is allowed to be expressed openly. Honest feedback improves our products, discussion breaks down barriers, and challenges to leadership keep us all on our toes. When senior leadership listens to line associates, the real issues of the organization come to the surface. We want our truth unvarnished, warts and all.

Here's what it looks like in practice: *Speaking up in meetings, asking tough questions, admitting ignorance, swallowing defensiveness, listening carefully, giving honest feedback, not whining, saying thank you, offering suggestions, stifling gossip, confronting conflict, questioning political correctness, rewarding courage, respecting others' points of view; open books, all-hands meetings.*

Harriet: Alan feels really strongly about this value. At a company meeting early on, an associate asked a question about a recent spike in workload. His team was getting swamped, and he was upset about it. One of the executive vice presidents gave a long, convoluted explanation. Instead of sitting back down, the associate pressed harder. "We had the same problem last year," he pointed out. "We should have foreseen it. What are you going to do about it, so we can avoid the same problem next year?"

There was a moment of stunned silence in the room. We'd never had so direct a challenge to our leadership before. Suddenly, Alan leaped off the stage, ran down the aisle, and gave the associate a big hug.

"That's just great!" he said. "That's just the kind of open and courageous communication we need. We screwed up, and you held us accountable."

The associate thought for a moment, then laughed.

"Wow, Alan," he said. "You always said we should embrace tough questions, but until now, I didn't know you meant literally."

2. Risk Taking

Harriet: We want everyone who works at Grand Circle to be able to lead at a moment's notice, and from anywhere in the organization, but that's only possible if everyone has daily practice taking risks. We encourage that by providing an environment where it is safe to take risks and make mistakes and where important risks are rewarded whether they succeed or not. Risk taking has a lot of benefits. It builds personal courage and corporate resilience, and, of course, it helps us capture opportunity. It has allowed our company to adapt to changing circumstances and grow to its full potential, even in the face of difficulty.

Here's what it looks like in practice: *Stepping outside your comfort zone, trying new things, moving forward without knowing the outcome, accepting new assignments gladly, seeking challenges, expecting mistakes, questioning assumptions, eschewing popularity, embracing challenges, catching others when they fall; constant measurement and evaluation, fast-exit plans.*

3. Thriving in Change

Alan: Change is a way of life at Grand Circle. It's part of our vision—to help change people's lives—and it's the nature of the travel business, which is highly volatile. Our goal for change is not just to survive, but to thrive, and so we practice. We change everything—all the time. We change products, business practices, organizations, priorities, and work assignments. Every change has its reason. Some are course corrections dictated by world events and global competition; others are issued as deliberate challenges to our associates, to help them grow into stronger leaders. In all cases, our goal is to use change to maximize our effectiveness and success in an unpredictable and ever-changing environment.

Here's what it looks like in practice: *Moving forward, turning on a dime, seizing opportunities, getting psyched, packing your bags, going out on a limb, embracing chaos, cutting your losses, letting go, staying positive, remaining calm, stepping up to help, being part of the solution, celebrating success; false starts, new horizons, no second-guessing, no regrets.*

After 9/11, we put "thriving in change" into play. The September attacks hit us very hard financially. Travel literally came to a standstill. In October, our bookings were down 90 percent; by November, cancellations were up 80 percent. Our vision and mission told us what to do: *keep traveling; change people's lives.* So we dug in, cut our product line by 20 percent, laid off 250 associates, and got some big concessions from our vendors. We needed customers, and so we offered a cancel-anytime, no-questions-asked travel policy—the first in the industry.

We stayed positive. Our vice chairman, a West Point graduate, gave the greatest pep talk in our history. He gave it at a corporate meeting, on a Thursday morning. "We're facing a major challenge, and you need to decide if you want to stay and help us fight it.

Take the weekend to decide, but decide. You're either on the bus or off the bus, but if you're on the bus, get ready for the ride of your life." On Monday, at the end of the weekend, every single associate returned, ready to ride.

By the end of the year, we had found our opportunity: as other travel companies shed their inventory, we bought it up, nailing down long-term contracts at discounted prices for flights, hotel rooms, meals, and other services that our competitors had previously locked up. When travel resumed, our competitors had to settle for second-tier properties at higher costs. We emerged as the dominant player in several new regions, and 2002 was our most profitable year on record. We trusted our vision, mission, and values, and they pulled us through.

4. Quality

Alan: This value comes directly from the argument on the Salmon River. We needed to make it plain: no more "cheap and nasty" vacations; high-quality unforgettable experiences would be the key to the Grand Circle brand. Most businesses believe cost and quality are constant trade-offs. We believe it's possible to drive improvement in both simultaneously, but if an irresolvable conflict arises, then quality always gets the nod. It's not just that we have high standards. Quality also drives our repeat business, which is critical to our financial performance.

Here's what it looks like in practice: *Reading every traveler survey, considering every suggested improvement, not settling, celebrating progress, knowing the competition, exceeding expectations, holding our own work to high standards, aiming for 100 percent.*

Harriet: Our travelers are mostly retired people. They have a lot life experience, and they love to talk. What a godsend for the company—our customers are our built-in quality control.

We currently have more than 115,000 people who travel with us every year, and we ask all of them to fill out a very detailed post-trip evaluation. As we noted in Chapter 1, we get about a 70 percent response rate—which means we receive about 80,000 surveys every year. That's a lot of information, and we welcome it. It is absolutely the most important information we can get about our trips—better than sales figures, better than trip leader reports, better than our own best judgment.

We love reading what our travelers have to say in their own words, and you'd be surprised by how many good ideas we've gleaned from these comments. Once a traveler on one of our early trips to Russia wrote, "No more god-damned potatoes!" Enough said. The food vendor was gone within the month.

5. Speed

Alan: You need to move fast in the travel industry, and so speed is one of our most important values. It keeps us alert and engaged, puts us ahead of our competition, and ensures the safety of our travelers in crisis situations. But not everyone is comfortable with speed. We help by creating a safety net, quickly changing direction if a decision proves wrong and ensuring that mistakes are reversed without penalty. Open and courageous communication also puts the brakes on speed. A wrong decision can be costly, but if individuals speak up when they see something amiss, it takes much of the risk out of speed.

Here's what it looks like in practice: *Feeling urgency, setting deadlines, beating deadlines, wearing a watch, answering e-mails immediately, seeking clarity instead of certainty, being proactive, urging teammates on, not lingering over drafts; fast meetings, full date books, the thrill of the chase.*

6. Teamwork

Harriet: Our values define our behavior only if they are shared by all of us and we work as a team to benefit both the traveler and the company. At Grand Circle, "teamwork" isn't a cheerleading slogan; it's the way we work on a daily basis.

Here's what it looks like in practice: *Sharing knowledge and expertise, stepping in to help, committing to a common goal, accepting responsibility, deferring to greater skill, providing honest feedback, challenging the team, pushing for better results, showing compassion, supporting other people personally and professionally, celebrating success.*

Put Your Values to Work—
Don't Just Hang Them on Your Wall

Alan: Our values are not idle words on a page; in fact, they get quite a workout. We talk about them all the time: in interviews, in meetings, in trainings, and in our newsletters. Every associate knows them by heart—not just in Boston, but in every office around the world. They are our collective conscience and our courage, and they literally govern how we behave. We use them to size up job candidates, to plan strategy, to make work assignments, to evaluate performance, to direct our growth, to measure our progress, and to decide who gets awards and recognition.

Harriet and I insist that our leaders exhibit the values constantly—beginning with the two of us. Everybody's got to "walk the talk." We stumble occasionally, but our associates and advisors always bring us back in line. That is what open and courageous communication does for us personally. It's invaluable. Our associates believe in our vision, mission, and values, and they trust us to keep them whole. They are disappointed if they see us fall away, and they remind us to remain true to our principles.

For example, late one Friday afternoon in 1994, after all the executives had left for the weekend, I sent out a company memo saying we were going to cut associates' bonuses because we had a big shortfall. My timing was terrible, and my delivery was callous. At the corporate meeting the following week, Anne Marie Davis, one of our account managers, let me have it.

"You can't just dump and run," she said. "If you've got bad news, you've got to stick around for the fallout. That's what 're-sponsibility to associates' means."

Anne Marie was right. Since that day 16 years ago, we've made sure to deliver bad news right away, honestly, and face-to-face in small-group meetings. We don't sugarcoat it, but we relate the news personally and explain exactly what we are going to do about it.

Anne Marie's feedback also resulted in another change that has become a signature piece of Grand Circle culture: our company report cards. Twice a year, each associate is asked to grade the company's performance on each of the four responsibilities—associate responsibility, customer responsibility, financial responsibility, and social responsibility—giving letter grades, A to F. On the first report card, our grade for financial responsibility was a D–. Anne Marie reminded us that leadership requires accountability, and now our report cards keep us honest.

Give Yourself the Freedom to Be Great

Alan: So, there it is: vision, mission, and values, the three elements of our corporate culture. We believe that by making our corporate culture explicit, we have given our strong, independent-minded company of leaders the framework—and the freedom—to do their best work. The culture simplifies everybody's job by making it clear to all: here's our dream, here's how we'll get there, and here are the values we'll count on along the way. Our company has been very successful through some harrowing times, and we believe the culture is what has seen us through.

To a great extent, we've succeeded in maintaining the maverick style of our early days. Certainly no one walking through the door on Congress Street would mistake us for an ordinary company. There is a life-size carved elephant in the lobby, and hundreds of postcards from customers are shellacked like wallpaper in the entranceway; you can hear world music playing and see nature and travel shows on the flat-screen TV. Upstairs, everyone is in motion. There is an air of expectancy and surprise, as if somebody, somewhere, might be up to something; as if a bagpipe player might suddenly start roaming the halls, a hot-dog-eating contest might take place, or the executives might throw an impromptu party on the loading dock—oh, yeah!—that's happened before.

At the end of the day, business is all about people—people working together to do something useful, maybe something great. So, let me ask you this: if your vision and values don't drive your enterprise, what will? Last quarter's profit and loss statement? Your competitor's latest product launch? The tweets of a disgruntled shareholder on Twitter? Reactive leadership is a dangerous game. It's like Whac-a-Mole, where as soon as you beat down one mole, others pop up—no good will come of it. But if you build a culture that will allow your company to adapt and thrive in your market, you will reap the rewards of truly visionary leadership.

Leadership from Anywhere (and Everywhere)

Develop leaders at every level to make your organization more adaptable, profitable, and successful.

Harriet: We have a saying at Grand Circle, "Our associates are number one." People love to correct us. "You me n, *your customers* are number one," they say, as if they have caught us in a slip of the tongue.

Alan loves this moment. He knows he has built the company on contrarian principles, and he is happy to challenge conventional wisdom.

"Let me tell you something," he always answers. "If you don't make your employees your absolute top priority, you won't *have* any customers."

Over the last 26 years, we've built the most extensive global organization in the travel industry: 34 overseas offices, with more than 2,200 people working with us in 60 countries, including about 700 office associates and 1,500 tour guides and ships' crews. We give our overseas associates a lot of autonomy and authority to run our business overseas even though our headquarters are thousands of miles away in Boston. We do this because we've learned that the best trips are delivered by local people, and the worst crises are avoided when local people are empowered to make decisions on the spot. We believe that a great company that can thrive in changing times needs to develop leadership capabilities in every associate.

Our associates are our greatest asset and our first responsibility—it says so right in our mission statement. When we take care of them, they take care of our business. It's as simple—and as complicated—as that.

A Different Kind of Company

Harriet: Even before we bought Grand Circle, we knew we wanted to run a different type of company, and one of the main things we wanted to do differently was to put our people first. It was one of my demands, actually. I am a "people person" by nature and a teacher by training; I believe everybody's work should be fulfilling. I told Alan back on the beach on Captiva Island that if we were going to throw our life savings into Grand Circle—along with our young family—then we had to make Grand Circle a really great place to work. No bosses, no exploitation, no secrets, no backstabbing. That's what the travel business was like in those days, and I was sick of it.

Alan completely agreed. Alan has always hated authority, and so there was no way a traditional employer-employee organizational structure could work for him. Besides, we had some people in mind whom we wanted to hire—smart, experienced, independent-minded people—and they would never put up with that kind of hierarchy. We needed to create a workplace where our associates could enjoy good company, pursue leadership opportunities, speak the truth without fear of reprisal, and find personal satisfaction—every day.

"Find a job you love, and you'll never have to work a day in your life." Alan would have coined that statement if Confucius hadn't said it first.

Leadership from Anywhere

Alan: Our decision to put associates first wasn't entirely altruistic. It also made good leadership sense. Some businesses may be best served by a top-down, rule-based leadership model, but international travel isn't one of them. It amazes me that so many companies tried to operate that way in the 1980s. Back then, the industry was split between independent travel agents—22,000 of them—and a handful of large, direct-marketing companies like Saga, based in the United Kingdom, and Olsen Travel World, based in Los Angeles. The big companies were hierarchically organized and highly corporate in management style, filled with drones at the bottom of the ladder and yes-men at the top. Leadership in those companies was largely a matter of issuing orders and not taking no for an answer.

That leadership model doesn't work in the travel industry, not in my experience. International travel is a very volatile business. Disruptions and crises occur with surprising regularity in one part of the world or another. When a volcano erupts in Iceland or a revolution breaks out in Egypt, our business is directly affected. We need action fast, and we need it locally. The company can't wait for someone from Boston to turn up and decide what to do. The comfort and safety of our customers and the survival of our business depend on responsive action by those closest to any emergent problem.

This means that everyone in the company must be willing and able to make decisions at any time, in any place, and at every level of the organization. We call this "Leadership from Anywhere," and it is asking a lot. It means that the moment you sign on to work at Grand Circle, no matter what your title or area of responsibility, you will be required to think and act and speak up as if the company depends on you, as if leadership could fall in your lap tomorrow—*because it can and it will.* At Grand Circle, every associate is a leader-in-waiting.

In the beginning, we hired people we already knew from our years in the travel business, people we were sure understood our vision for the company. Most entrepreneurs and takeover teams do this, and it makes sense. So does asking those same people to refer like-minded friends and associates; it makes the early months of the enterprise run more smoothly. Even today, 38 percent of our new hires are referrals—a high percentage compared with that of other businesses. That's a strong core of people predisposed to understand Leadership from Anywhere and to thrive in our corporate culture.

Of course, not every job candidate comes to us with this kind of leadership thinking. It is our responsibility to develop leadership at all levels of the company by giving all our people opportunities to lead from wherever they are, to become the best leaders they can be. That's why we put so much emphasis on open and courageous communication, risk taking, speed, teamwork, and the ability to thrive in change. These are values that build leadership skills, and they are values we expect all our associates to embrace—beginning with our very first encounter, in the hiring interview.

Hire for Values First, Skills Second

Harriet: Our hiring interviews are legendary in the travel business. Maybe *notorious* is the better word. Our interviews are part encounter group, part psych experiment: we design them to determine how well candidates will fare in our corporate culture. You either love the experience or hate it. I'd say the breakdown is about 70–30.

Our interviews are unusual because we hire for values first, skills and experience second. That's not the usual procedure, of course. Most companies start with a pile of résumés, looking for the candidates who went to the best colleges or who have the most suitable skills or experience; then they interview those candidates to see if they'll "fit in." We think that's backward. Skills can be taught fairly easily, but values take a lot of work. If speed, risk taking, courage, honesty, and teamwork are missing in a candidate, he or she won't make a good associate no matter how strong the résumé.

I once hired a woman named Martha Prybylo to help me run our charitable arm, Grand Circle Foundation, even though she had no foundation experience at all. She was referred by our head of public relations, who praised her public service work with senior centers and councils on aging. I didn't even ask for her résumé before the interview, because I could see she had the right priorities. She got the job, and she was so successful that we asked her to take over our human resources department 5 years later. She didn't have any HR experience either, but she excelled at that, too. She then became the executive vice president of our Worldwide People & Culture department (our version of HR) and the very captain of our culture and values for more than 17 years.

Consider the Power of the Group Interview

Alan: I think what flusters some of our job applicants is that we start them off in a group interview, and it's very interactive. We bring in four to eight candidates at a time to meet with two facilitators: someone from our People & Culture department and someone from outside People & Culture who has been with the company for at least five years and has a deep understanding of our values.

We tell the candidates in advance that they will be participating in a group interview, but we don't tell them whether they're all interviewing for the same position; sometimes they are and sometimes they aren't. The uncertainty creates some anxiety and competition, and we are interested to see how the applicants will handle it. After all, ours is a really high-pressure business. Will they hang back or jump in?

In fact, the whole 90-minute interview is designed to screen for people's willingness to engage and take risks. After introductions and a brief overview of the company, we present a series of team-building exercises and role-playing scenarios to see whether applicants can track problems and contribute to their resolution. One exercise involves an egg; another requires participants to pull challenging questions out of a bowl and answer them on the spot. For example, "What is the toughest feedback you've given to your boss . . . and what happened?" or "If you were CEO of your current company, what's the number one action you would take today?" The tasks change all the time, and so candidates can't prepare in advance. We're not so interested in how well the candidates complete the tasks we've assigned them as in how productively they interact with one another, how creatively they think, and how well they receive feedback.

Harriet: As a teacher, I always find the results interesting. Often one domineering candidate will try to take over an entire activity, while others barely participate at all—just like in a middle school classroom. Some people become uncomfortable and ask to be excused; others become furious and storm out. We're sorry we've made these applicants uncomfortable, but we're not sorry to see them go; and we never give second chances, because we know these people would never fit in with our culture. The more successful candidates, on the other hand, often feel that they have accomplished something extraordinary. They've been judged not on their résumé points but on their native abilities, their understanding, their true selves. They've already become invested in the company.

We developed this hiring process about 10 years ago because we felt our turnover rate was too high. We had talented and skilled people working for us, but they weren't staying because they didn't really fit in; there wasn't enough in it for them, or for us. Once we changed our interview process, our turnover rate dropped significantly: currently, it's only 6.6 percent, which we think is a pretty good number. The key to our success is the motivation and commitment of our associates, and that's the result of making our people our top priority. When people want to be part of a company, employee engagement is easy.

Our approach to interviewing might not work in your company. But do yourself a favor. Think about your vision, your mission, and your values; then think about how you recruit, interview, and hire people. Is there any connection there at all? Give the members of your HR team a day away from the office to brainstorm ways to put your company values on the hiring table in a really productive way. You might be surprised by what they come back with.

Not All Hires Can Be Successes

Alan: Not all our hires are success stories. Over the years, my biggest mistakes have been staying with people too long because they were high performers, even when I knew they didn't really share our values. The Salmon River off-site should have taught me better, but the lure of the big-shot high producer is strong.

There was this big guy—around 6 feet 7 inches and 270 pounds—whom we hired to be CFO in the mid-1990s. He had a lot of experience in the travel industry and a strong financial background. We were so impressed by his size and his skill set that we never took the time to check him against our values. As it turned out, he couldn't handle the off-sites; a tiny bit of drizzle sent him packing. He just wasn't a risk taker (though to his credit, he did appear in a lederhosen costume during the company Halloween party), and we soon parted ways.

Another time, we hired a former Disney exec to be our COO. He had such a big name and looked so good on paper that we bypassed our usual interview processes. He turned out to be arrogant and self-centered, not at all a team player. We should have realized he wasn't going to fit with our culture when he had to be physically restrained from attacking some of our associates when they put on a skit gently mocking Mickey Mouse and his corporate pals, but we let it go. You'd think such a hotshot would be good under pressure, but this man failed us on September 11, 2001, when the terrorist attacks took our company to the brink of ruin. I was clear across the country on a mountain-climbing trip with Mark Frevert, and I needed the top people in Boston to be calm and collected, but this guy completely fell apart and had to be driven home, leaving Harriet and Joe Cali, our executive vice president of marketing, to handle the crisis alone.

We learned our lesson. If you have a vision and a mission you believe in, you have to evaluate your people against them all the time, no matter how much you admire their energy or reputation and no matter how much money they are bringing in. If you don't, you will live to regret it.

Learning the Job

Harriet: We have a couple of strong training programs in place for our new hires. What's unusual about the programs is that, like the hiring interview, they are centered on values, not job skills. For example, all our new hires go through our Odyssey Program, a two-day orientation program. The first day we put everyone in a van and take them to Pinnacle, our outdoor leadership center in New Hampshire, where they do the high-ropes course and other team-building and risk-taking activities. The next day, they're back in Boston for a day devoted to the company's history, vision, mission, and values.

Call center hires get an additional four weeks of training, so they can learn all our trips and really understand what we mean by our mission of customer responsibility. Even after they've gone to work in the "Nest" (which is what we call our call center), they receive individualized coaching and help for another 60 days, to be sure they can apply our culture and values to the very sensitive work of customer relations.

Learning by Doing

Harriet: We are big believers in experiential learning at Grand Circle; we think the absolute best way to learn is by doing. This is Alan's personal legacy. Alan was never much of a classroom student. He got his early leadership education on the streets; later, he learned from some unusual people he met in his travels—people such as Sir Edmund Hillary, the Everest climber, and our friend Kipuloli Napiteeng, a Maasai tribal chairman we met in Kenya. Alan sees business and leadership learning as a lifelong project, and we take the same approach with our associates.

One way we foster "learning by doing" is by periodically giving associates new job assignments that change their work group and stretch their skills. A marketing copywriter might begin the day writing itineraries for a South Asia catalog and end it answering travelers' questions on our Web site. You never know what the day will bring—that's one of the wonderful things about working here.

We also make a point of transferring associates among departments. A transfer is a lateral move designed to give a promising associate experience in a new area of the company's operations. That's what I was thinking when I moved Martha Prybylo from the Grand Circle Foundation to the HR department. If she had worked for a more hierarchical company, Martha might have resented it. In those companies, corporate advancement usually comes with increasing competence and responsibility in a *narrow* area of expertise. But the price of this kind of advancement is high, both for the associate and for the company. When an associate is trained only to the specialized knowledge of a small segment of the company's operations, he or she can never become a true leader of the enterprise and can never realize his or her potential.

Learning from Everyone

Alan: You don't need to move people *permanently* to get the benefits of fresh ideas and teamwork. Sometimes you just have to bring different people together and encourage them to think outside their traditional job descriptions. We once held a leadership meeting in Morocco. We wanted to brainstorm ideas for creating really unforgettable experiences for our customers, with the goal of improving our top-20 trips all over the world. Associates from every key region and department came to the off-site, and we needed to break into teams.

Right away there was disagreement about what to do with the finance guys. Nice people, all of them, but they were a bunch of number crunchers, and they didn't know much about the trips. Most people wanted to distribute them among the teams, to spread out the inexperience. This is our usual approach to team building, but this time I decided to put all the finance guys on the *same* team. I thought maybe the geeks had something to teach us, and—sure enough—they came back with the best recommendations for how to improve the trips.

One recommendation increased our travelers' satisfaction with their shopping experiences on our trips, while simultaneously increasing our profitability from the goods that were sold. We take our travelers to shops that sell local goods or crafts that Americans tend to like—such as rugs in Turkey and India and pottery in Thailand and Portugal—and we get a percentage of their purchases. The finance guys noticed that travelers on our Morocco trips weren't rating the shops there very highly, and so they suggested we improve the shopping experience there. At first, we worked with the shops we had been working with, but in the end, we moved our business to several better shops that sold higher-quality goods. That increased the excellence ratings by our travelers, and since sales increased, so did our profitability. Kudos to our finance guys: they found freedom in their "outsider" position and strength in their numbers. That meeting led to terrific improvements in our trips, and we've continued to call on associates from all across the company for their perspectives ever since.

I believe our associates can lead from anywhere. And I know we always get great ideas when we give associates the opportunity to get away from their daily routine, to think and work outside their area of expertise.

What do you think your people could do if you gave them the chance to step outside their boxes? What if you randomly paired every person in the office with another employee—say, alphabetically—and asked each to give the other some suggestions about how they might do their job for the company differently? Could your mailroom guy suggest cost efficiencies to the CFO? Could your head mechanic tell your receptionist something about how to answer customer calls? Give rewards for the best team suggestions. What's the worst that could happen? You could lose a couple of hours' time and maybe some prize money. But you will empower all your associates, and you might just learn something interesting.

Transformation and Adaptability

Harriet: Another way we share associates' experience is by forming cross-functional teams. We do this whenever we need to resolve big issues or change the way we do business. We call them "transformation teams," and they have ushered in some of our biggest changes, like the opening of our first overseas office, in Munich, and our purchase of Overseas Adventure Travel. Assignment to one of these transformation teams is an honor, and it means a lot of work; team members must quickly learn the needs of all departments and come together, through a group leadership plan, to move the company forward.

This kind of cross-departmental training is a huge help during crises because we can count on many associates bringing firsthand experience to the crisis room. For example, in October 1999, we experienced a terrible tragedy when EgyptAir Flight 990 crashed off Nantucket in the middle of the night, killing 54 of our travelers. Family members called us for days, first to get information, later to arrange transportation to Boston, and finally to get help with memorial plans. All our associates wanted to help. In fact, when I first got to the office, just hours after we got the news of the crash, we had more volunteers for the call center than we had phones available. Because of our policy of cross-training, most of those volunteers had call center experience and could step right into the job. Similarly, two years later, after 9/11 disrupted air travel all over the world, we were able to call on many associates with air-routing experience to help us get our travelers home.

Your company's crises may not be as heartrending as ours—I certainly hope they're not—but you can be sure they will be similarly challenging. You might lose a big client and not know how you're going to make up the shortfall in revenues. You might lose one of your top people to illness or sudden death and need to handle all the responsibilities that person managed so well. You might face a product recall, or a public relations scandal, or a production breakdown, or a deadline crisis. But if you have cross-trained your associates to handle myriad functions, they'll help lead you out of the crisis from wherever they are. Your employees are your best assets, but only if you give them the values and training they need.

Leadership Takes Practice

Alan: It isn't enough to say "Associates are number one" or "We believe in Leadership from Anywhere." You've got to make it happen. And we do, all the time. Our associates may be called on anywhere and anytime—in a training session, at the lunch table, or on a site visit—to tell us what they see as the company's hot issues and how they would help us improve our performance.

I'm famous for confronting new hires in the elevator, shortly after their first company meeting.

"How'd it go?" I always ask.

"Great!" they always say.

Then I ask, "What are the top three things you would do to make our meetings more effective?"

The new hires are usually surprised by this, and tongue-tied. It takes a while for new associates to understand that we *really* value their opinion and that we *really* expect them to give us feedback so that we can continually improve the business. It makes some people nervous, at least in the beginning, but like public speaking, leadership gets easier with practice—and we practice all the time. We are fixated on empowerment and readiness. We call it our "culture of leaders," and we keep telling our associates, "You can lead from anywhere."

We would like every person in the company to be a leader. That may be impractical, and not everyone rises to the challenge, but we are committed to helping people all over the world become the best leaders they can be.

Earlier in this chapter, Harriet mentioned our friend Kipuloli Napiteeng, a Maasai tribal chairman who is an outstanding—and colorful—leader. Before we go much farther down the leadership road, let me tell you how we met him and got to know him. In January 2001, Harriet and I traveled to Tanzania, where our friend and partner Willy Chambulo, owner of Kibo Guides, arranged for us to camp at a Maasai village called Sinya. The day we arrived was market day, and streams of people were coming to market from all directions, bringing sugar and trinkets. They were dressed in the traditional red cloth of the Maasai, and their spears were rattling and their jewelry tinkling.

Somehow we found ourselves in a grassy area under a grove of trees where the villagers were slaughtering a goat. Pretty soon, we were attracting a crowd. At one point, Harriet was surrounded by throngs of chattering women dressed in boldly printed fabrics, all eager to see and touch this white-skinned, tawny-headed visitor. They were smiling, singing in a high-pitched way, and ducking their heads toward her and away from her. It was like being caught up in a raucous celebration of some unexpected victory. The warmth of the people and their eagerness to share their culture made this truly an unforgettable experience.

But it was also a little scary. The crowd was too close, and Harriet couldn't get any air. Suddenly, a man appeared, carrying a spear, and the crowd dispersed. Something in Harriet's expression had persuaded him to intervene. This was Kipuloli Napiteeng, the head of the village, and for the rest of our stay, we were Kipuloli's honored guests.

That evening, Kipuloli, Harriet, and I shared a meal around a campfire. We had a long and fascinating conversation about our respective lives. We learned that Kipuloli had been elected chairman by the male elders of the village and that he shared our passion for education. He understood that education would irrevocably change his village and its culture, but he wanted the children to make their way and prosper in a modern world. We saw that Kipuloli was a compassionate and visionary leader, and we liked him immediately. We didn't want this unusual cultural exchange to end, and so we made an impromptu decision: we invited him to come to Boston.

That's when we found out that Kipuloli possessed another leadership skill: he knew how to bargain. He would come to Boston if we would meet three conditions: that he could bring along his chief warrior, that they could both wear their traditional dress, and that they could carry their spears (except on the airplane, of course). We agreed, and Kipuloli agreed to our sole condition—that he keep an open mind. The deal was made: he would come in June.

What were we thinking? Kipuloli had never been on an airplane, and he had seldom been more than a few dozen kilometers from his village. He spoke no English, and we spoke no Maasai; Willy had been our interpreter during our entire visit. Kipuloli had never seen a building taller than a few stories, and he had only a rudimentary understanding of money. How would we even feed him and his warrior? Here, we were eating goat, which Kipuloli had brought and cooked over a fire—but there are no Maasai restaurants in Boston!

We shouldn't have worried; when he came, he fit in immediately.

That June, he joined us for the conclusion of BusinessWorks, our once-a-year company event that brings together associates from all over the world. We had been meeting for two weeks with our Boston and overseas leaders to resolve business issues at Pinnacle, our leadership development center in New Hampshire. On the final day of the off-site, the entire company came together for a day of learning, challenge, and some irreverent fun. Kipuloli and his chief warrior were waiting with us as the buses arrived with all our Boston associates.

Although Kipuloli didn't speak English, he didn't need to be told this was a tribal gathering. He is a great leader, and he knew just what to do. Kipuloli and his warrior began to chant; then they started leaping into the air in time with the rhythm of their chant. Within seconds, scores of people, including Harriet and me, were leaping into the air. A few of us even tried to join in the chant. Something was happening—something spontaneous, foreign, and yet completely in keeping with the Grand Circle culture. For 15 minutes, pandemonium consumed the previously quiet meadow. The Maasai chairman and his warrior had never been out of Africa and had never flown in an airplane. But they sure knew how to jump-start a gathering of the clan. It was impulsive. It was raucous. It was an unforgettable experience. Kipuloli led us that morning, and we followed his lead.

Three Essential Qualities of Leadership

Alan: It always makes me laugh when some self-important CEO gives a talk about the "mysterious quality of leadership," as if leadership were some divine attribute possessed only by a chosen few, like second sight or flaming red hair. That's a crock. Leadership isn't mysterious, and it's not rocket science—as we saw when Kipuloli jump-started our off-site meeting. I think leadership boils down to just three things.

First, a leader must have a vision and be able to communicate it effectively. If a leader doesn't know where he or she is going, who's going to follow that leader? A leader must also see the world realistically and approach problems pragmatically, so he or she will know what it takes to actually achieve the vision.

Second, leaders must be gutsy. *Gutsy* is Harriet's word, and it means leaders must tackle tough issues and make decisions about them right away—no shirking. Gutsy leadership takes courage, but it also requires humility, because you have to allow for mistakes. Gutsy does *not* mean fearless! In fact, fearless people make terrible leaders, because they can't be depended on. Sooner or later, a fearless leader will do something reckless and irresponsible—and take the company down.

Third, good leaders empower others to help them achieve their vision. They celebrate other people's successes, and they give back to the people and places that support them in ways that help others grow into their own potential.

Leadership like this is in short supply, but the good news is that it can be taught. You'll have to confront a lot of follower behavior first, however, because people all over the world have been trained from childhood to defer to their elders and bosses and to conform to the supposed needs of the larger organization. We've learned that when people are encouraged to act as leaders instead, more than a few jump at the chance. They respond with joy and enthusiasm, like it's some repressed need.

Age doesn't seem to matter; a person can rise to leadership at a young age, or in old age, or at any time in between. Gender is completely irrelevant in our experience. Titles and position don't matter either; we've had associates step up to leadership from the Nest and boardroom equally. What counts is decision making and action—resolute action that is guided by vision, undertaken courageously, and directed toward the greater good of the company and the world at large. Understood this way, leadership can truly come from anywhere.

WHAT GOOD LEADERS DO

> Hold a vision
> Communicate effectively
> See the world realistically
> Listen
> Deal with the tough issues first
> Make courageous decisions
> Recognize mistakes and reverse bad decisions
> Develop other leaders
> Recognize and celebrate success
> Give back

Embrace the Hot Seat

Harriet: One way to understand our leadership model is to sit in on one of our corporate meetings, which we hold once a month in the big open space on the fourth floor of the Boston office. These are all-hands meetings—all 475 Boston associates (except those call center associates not scheduled to be in), along with any overseas associates who are in town on business. They are high-energy events, beginning with the rock music that our associate Ricky

Regan plays to welcome the crowd and ending with Alan on the "hot seat."

We open the meeting by welcoming new associates and honoring those who have reached milestone years with the company. Ten-year associates are expected to say a few words about how Grand Circle has changed their lives. One associate recently explained how a trip to Egypt helped her overcome a lifelong fear and suspicion of Muslims, a cultural barrier she had grown up with in her native Philippines; another spoke movingly of how open and courageous communication helped her get past a troubled domestic relationship. We hear stories like this all the time, and they are gratifying.

Next come excellence and teamwork awards for outstanding performance. We expect excellence and reward it, and we ask associates to recommend their peers. Prizes include cash, recognition in our weekly e-newsletter, and a free overseas trip known as the "recognition journey" for all of the year's winners the following spring.

Next we produce the month's profit and loss statement, a summary sales report, the quality scores from travelers' evaluations, and an accounting of all our mistakes. Full disclosure, nothing held back. This kind of open-books policy is practically unheard of in privately held companies like ours. We do it to honor open and courageous communication and to make sure that everyone is aligned on the same issues and goals. Besides, how can we expect leadership from all levels if we withhold vital information?

The corporate meeting goes for an hour. The second half brings the fireworks. That's when Alan and the rest of the executive team sit on the hot seat while associates ask them uncensored questions. The tougher, the hotter, the better. They have to answer all the questions, and it can get intense. You can feel the apprehension when an associate lobs a challenge to the front of the room, but you can also feel the empowerment that swells up when other associates rise to their feet to applaud the question that has been on everyone's mind. *Why is the incentive plan a month late? Why aren't there more associate travel deals offered this summer? When will the 2012 products be ready—customers are complaining!*

We have trained our associates well. They don't tolerate equivocation, and that's just as we want it. We're looking for honest feedback, not blind loyalty. Alan always says, "A yes-man is not a leader. He's not even a good follower. He's just taking up space."

Our corporate meetings serve many purposes. They keep everyone informed. They promote communication across all levels and departments. They remind our senior executives to "walk the talk" by exhibiting our values in public and productive ways. And they empower associates to lead. They are like old-fashioned New England town meetings: sometimes painful, often raucous, and always productive.

Think about what open meetings and full disclosure could do for your business. If your employees knew more about how their work contributed to your goals, how much more successful could your company be?

Go Off-Site to Get on Track

Alan: I think the best leadership work happens out of the office, and so I take the executive team on an off-site whenever we have a tough issue to resolve. For example, I took a team to Spain in 1996 to brainstorm the rollout of our first overseas offices, and I took a team to Cuba a few years later to figure out how to build a stronger worldwide organization. Typically, we have three to four executive off-sites a year, sometimes in places like the Turks and Caicos and other times as close as Newport, Rhode Island. They are incredibly powerful. Something about the change of scenery opens up thinking and generates new ideas.

A lot can go wrong, of course. We learned that on our very first off-site, when we went white-water rafting on the Salmon River in Idaho. Between the rough water and the arguing, we were lucky to come back alive. That trip taught us to dial back the physical challenge, tackle the toughest issues first, and be ready to channel emotion. Now we keep our discussions on a direct but respectful footing, focusing on learning, open communication, teamwork, and company advancement—not conflict. Our off-sites are tamer now, but much more productive.

Whatever the problem we're tackling, we make a deliberate effort to develop the values of open and courageous communication and risk taking. I make sure the team owns the issues and results—I don't hand them down from the top—and during feedback sessions, the participants come up with their own professional development plans, which they execute when they get back to the office.

One of the great things about owning a travel company is that we can offer leadership training in really exotic locations. We can literally use the world as our classroom, even for associates who are not members of the executive team. For example, travel figures prominently in our signature Leadership, Exploration, Adventure, and Discovery (LEAD) program, which takes a group of high-performing associates from all levels of the company on a 10-day version of one of our trips. The second day of the trip usually includes a challenging team-building activity designed to get individuals out of their comfort zones. LEAD groups have slept in caves, hiked through mountains, camped in the Arctic Circle, taken part in a road rally in Egypt, and gone bungee-jumping in New Zealand.

Although the focus of LEAD trips is leadership development, there is also a business component. All participants are expected to return to the office with recommendations for improving the quality and profitability of the trip.

Breaking Down Barriers to Build Leaders

Harriet: In the early 1990s, Alan began taking the entire Boston office on a summer off-site to New Hampshire. At first it was a two-day affair, based on an Outward Bound model. We brain-stormed ideas, slept in tents, climbed telephone poles, walked high wires, and swung from ropes. Not everybody loved it, but it was a good way to practice our values, especially risk taking, teamwork, and open and courageous communication.

Today the program, called BusinessWorks, is our most power-ful leadership development program. What began with about 100 Boston associates has grown to include some 400 associates, pri-marily non–call center folks from Boston along with leaders and top guides from our overseas offices. We still camp out and do the ropes courses, and we put on some pretty uproarious skits, but we also tackle some serious business issues. It's amazing to watch a program director from Thailand argue with a purchasing agent from Egypt over vendor delivery for a trip to Mongolia—all while sharing a watermelon in the New England woods! It's a cross-cultural exchange that takes company learning to a whole new level.

Alan: BusinessWorks is all about breaking down barriers and leading from anywhere. Many of the people who come to Busi-nessWorks are already company leaders. The executive team always comes from Boston, and so do many managers from our overseas offices. But we also invite people who are not in leaderships posi-tions, including high-performing associates, high-scoring guides, some of our grant recipients—sometimes a whole lot of travelers.

The idea is to focus a lot of talent and experience on our busi-ness goals. We do that by giving the group some outrageous but timely challenges. One year, we challenged participants to find ways to save $52 million in costs; another year, we asked how we could appeal to our women travelers, who make up two-thirds of our business. This business portion of the off-site can last one to two weeks, and every participant leaves with actionable and mea-surable commitments to put into play right away.

It's hard work, and it can take some leadership on the fly from me, but the outcomes have been amazing. For example, I once challenged participants to improve our "discovery scores" by 10 percent across the board. Our discovery scores are part of our travelers' trip evaluations: they measure how well we connected our travelers with local people, cultures, and customs. *How cool was the experience? How authentic?* When I put the challenge out there, I was met by silence. The guides were afraid to risk any innovations that might lower their *own* performance scores—scores that determined their incentive pay and the number of trips they could lead.

So I gave them all "free passes."

If the guides wanted to try something new and unusual on a trip, we wouldn't count that departure in their performance scores. Suddenly the ideas came pouring in—a visit to the red-light district in Sydney, a ride on the lavish Blue Train in Serbia, a meeting with a caravan of Tinkers (Gypsies) in Ireland, a sampling of rat in Thailand (optional!). Sure enough, the free-pass departures outscored the regular departures by more than 10 points in the field tests. Some of the excursions, like a visit to the City of the Dead in Cairo, were so popular that we made them standard offerings.

Did all the experiments work? No, not all of them. We measured their success from traveler feedback, kept the best ones, made changes to some others, and dropped the ones that bombed. The whole process exemplifies our approach to business: tackling tough issues, listening to our associates, empowering them to make decisions, reversing mistakes without consequences, and celebrating success. In this kind of business climate, associates really can lead from anywhere.

Action Learning Shows How to Work as a Team

Harriet: On Friday, the last day of BusinessWorks, participants are joined by most of the Boston staff for a day of group learning, challenge, and fun. In some years, we've even shut down the call center. This day is widely known around Congress Street as "the day we love to hate." First, there's the insanely early wake-up call.

By 7 a.m., associates are already on the bus, heading north and fretting about having to climb up high ropes or to sing in front of 400 of their peers during the afternoon skits.

Next thing you know, we're flying through the trees on the high-ropes course and building rafts from loose barrels and lengths of rope. The challenge is to construct a raft that can support six people and paddle it across a pond. We work as departmental teams, and by midday, we're beginning to appreciate unsuspected qualities in people we've worked with for years. The quiet guy in accounting knows everything about knots; the head of the marketing department can't swim. Somehow, everyone makes it to shore. We call it "action learning," and it teaches us a lot about how we set goals, handle conflict, rein in the bossy teammates, buck up scared ones, and persevere together until we get the job done. We'll bring these new understandings back to the office, where we can use them to improve our teamwork.

In the afternoon, everyone gathers under a giant tent to put on skits. There is rollicking laughter and spirited balloting for the best ones. Irreverence toward senior leadership gets huge points from the crowd. Bad singing and cross-dressing are often involved.

All these experiences—the intense business work, the physical challenges, and the unbridled fun—underscore our values of risk taking, teamwork, and open and courageous communication. They strengthen our corporate culture and contribute to leadership development by affirming the principle that leadership comes from every level of the company. Associates find out what they are capable of at BusinessWorks, and the experience can change their lives. An associate once told me it took her four tries to climb up the tree to reach the high wire that one can then walk across, but she finally did it. Many associates scramble quickly up the tree, hop on the wire, and traverse it, but she could not. She was terrified of heights. So while she did not make it along the high wire, she was able to conquer her fear of climbing the tree, and for that, she was the hero of the day.

High ropes and leadership roasts can be hard sells in traditional companies. Are your employees brave enough? Are *you*?

Create Your Own Leadership Training Center

Alan: BusinessWorks off-sites are held at our Grand Circle Leadership Center, a 500-acre site in Kensington, New Hampshire, that we own and operate. We use it for our own leadership training and for fee-based leadership development programs for corporations. We also offer its use to social entrepreneurs and not-for-profit organizations free of charge.

Grand Circle Leadership is run by professional organizational advisors who offer a comprehensive business and leadership curriculum. The facility has cabins and studios for indoor meetings, several different venues for outdoor meetings, a network of trails, several fields for orienteering and team problem-solving exercises, a huge pond, and an eight-station ropes course in the treetops.

Over the years, the leadership center and Grand Circle have contributed significantly to each other's success. The center's staff has developed many business and leadership training modules for us, including single-day and multiday sessions on such topics as "Building a Common Worldwide Organizational Culture" and "How to Turn Trouble to Opportunity"—all using leadership principles and business models that have made Grand Circle so successful. The center has trained scores of leaders for Grand Circle from countries around the world, including staff from all our overseas offices.

At the leadership center, we use a methodology called the *nominal group process* to help us examine and resolve tough issues. The methodology is simple. Breakout teams or buddy pairs write their top concerns and ideas on flip-chart paper. When the team regroups, all the lists are compared. After eliminating redundancies, everyone votes to prioritize the issues. The top vote-getters are put in rank order, and these become the issues and concerns that will be resolved. The same process is then used to come up with the solutions and action sets.

We also use this process every day in our offices and on all our company off-sites, wherever they are held. In all cases, an associate acts as facilitator, ensuring that everyone has a chance to speak, that the other participants listen and ask only clarifying questions, and that there is consensus on the rank order of issues. This is a very different approach to problem solving than coming in with an agenda and predetermined talking points. It raises everyone's concerns and ideas, allowing a consensual leadership to emerge from all levels of the organization, whose members then own the issues and actions.

Leadership Is Not for Spectators

Harriet: Good leadership is the force that holds a company together and simultaneously moves it forward. Good leaders meet new situations head-on. They inspire others to action. They jumpstart new initiatives and find opportunities wherever they go. We love that kind of leadership, that willingness to engage right then in the moment, to create something from nothing, to keep the engine running.

Over the years, we have seen just how powerful that kind of leadership can be. We see it in our guides, whose leadership skills make our trips so memorable. We see it in our leadership team in Boston, which has courageously navigated us through crisis after crisis. We see it in our best overseas managers, whose trips get the best scores from our travelers. We see it in our Grand Circle Foundation projects, which flounder without strong local leadership. We believe to the depth of our being that we need good leaders to help change people's lives. Leadership is important at every level of our enterprise.

More than anything else, leadership means engagement in the here and now. We love the fact that Grand Circle associates are engaged with customers, their teammates, and the community. What makes our business fun is watching people stepping forward with great ideas, ignoring organization charts to challenge authority, demanding to have ever-greater challenges, and jumping into the middle of a crisis to help. We are rarely disappointed and often astounded by our associates.

"Our associates are number one"—that's our first Extreme Competitive Advantage. We work with our associates every day to cultivate a corporate culture that propels performance and gives everyone opportunities for growth and leadership experience. If our culture delivers what our associates need, then we feel confident that our associates will deliver what our customers want. By empowering our associates to lead from everywhere, we guarantee the success of our enterprise.

In today's cost-cutting corporate climate, employees are too often regarded as liabilities—resources to shed if there is a way to get the work done more cheaply in China or India. We see our employees very differently; to us, today's associates are tomorrow's leaders. And we don't mean some tomorrow in the distant future. We mean literally *tomorrow*. We know that our associates will step into any leadership vacuum whenever they are needed and that they will help lead the company to ever-greater achievements.

What would your associates do for your company if you gave them the power to lead?

In times of change, an organization that focuses on developing its associates and building a culture of leadership from everywhere will be able to adapt and prosper. Organizations that depend on traditional structures will resist adaptation and will eventually be left behind.

Doing Well by Doing Good: Integrating Philanthropy with Business Strategy

Make philanthropy part of your business strategy: to do well and do good.

Harriet: Philanthropy is not a cultural impulse everywhere, especially not in the corporate boardroom. Most CEOs give only cursory attention to the idea of giving back. Their charitable efforts are minimal compared with what they might be able to do. In fact, we admire only a handful of companies for their philanthropy: Microsoft and Timberland, to name two. Maybe most CEOs think charity has little or nothing to do with business. That's not how we think, and it's not how we run our business. We believe philanthropy doesn't *hamper* business; instead, it *enhances* it. We believe philanthropy is part and parcel to building a great company and should be a key business strategy of every CEO, regardless of its revenues or size.

Philanthropy is good for business for so many reasons. It attracts good people to your company. It helps motivate and retain those people. It creates a personal connection with customers. It builds trust in your organization. Both associates and customers feel emotionally attached to an organization that is doing good things and not just making money. That attachment benefits your company in many ways, especially when you face challenges.

Philanthropy has become part of our corporate culture, and it's an integral part of our trips. It's not an isolated act of long-distance charity, but a vibrant part of our day-to-day business and a grateful expression of our global citizenship. When I think of global citizenship, I always describe it as a "tapestry": we're here in this world by ourselves, individually, but when you thread us all together, we're really something spectacular. We've found this to be true at Grand Circle, and we hope what you read here will give you some ideas for how you might integrate philanthropic endeavors into your business.

Commit to Philanthropy as a Company (Not Just as a Bunch of Individuals)

Harriet: When we took over Grand Circle in 1985, we didn't give much thought to *corporate* philanthropy. We were too busy turning around the company so that it would stop losing money and remain viable. We did donate our own time and money to various charities and community service projects. We were grateful for the opportunities we had had in life, and we felt a responsibility to help others. Long before we bought Grand Circle, we gave money to charities like the Shriners to help sick children. We gave to Neurofibromatosis, Inc., for research to cure this devastating neurological disease. We gave to Thompson Island Outward Bound to help provide outdoor leadership training for inner-city youth in Boston. We supported many local causes and organizations, including our neighborhood YMCA. We supported a holiday tree-lighting project on Boston's Commonwealth Avenue mall, a block from our home.

We also thought it was important for us to show up in person, to give our own time to local causes. We believed our personal involvement meant the money we donated would be used more effectively.

We also wanted to share our ideas about personal giving with our children. So every year around the winter holidays, we took Edward and Charlotte to help out at the Boston Family Shelter. They were young then, barely in school, and they loved all the excitement around the presents and the Christmas tree. Perhaps they didn't quite grasp the problem of homelessness, but it was a start.

We also promoted the idea of "giving back" at our fledgling company. We encouraged our associates to donate their time to community initiatives in Boston and their hometowns. And we supported charities our associates asked us to help, as a way to encourage their involvement.

But until 1992, our efforts at philanthropy were really scattered. Although we included the goal to help change people's lives in Grand Circle's vision statement, we didn't have a formal structure in place to make it happen. This was soon to change.

Alan: In February 1992, we went on an 11-day vision quest in Arizona, just north of Prescott. We didn't do this on a lark; we went because we wanted to discover more about ourselves, work on our personal relationship, and recommit to helping people. We had high hopes, but we never guessed that this would be the most powerful experience of our lives, or that this program would so wonderfully enhance our vision. Twenty-eight people participated in this unusual adventure, but Harriet and I were isolated from them for 3 days and 2 nights—with no food, no tent, no radio, and no phone. All we had were sleeping bags and water, and we were not allowed to talk at all during this time. During the days, we walked (sometimes together, sometimes alone) and thought about our lives; at night, we stared at the starry sky, which inspired wonder and introspection. We felt a strong connection with the world and beyond.

On the third night, I woke up at three in the morning with a revelation: To really help change people's lives, we needed to make a stronger commitment; we had to use the entire company to make a real difference to our travelers and the places we visit. This idea—a Grand Circle charitable foundation—would grow through the years to help countless people, but it was born in the still of the night with not another soul around us for miles. For us, the vision quest was exceptionally powerful. Neither of us has felt anything like it before or since. It was life-changing. It didn't alter our path, but it showed us how to accelerate our pace toward the goal of helping to change people's lives.

Of course, we're not suggesting that *you* need to go on a vision quest in order to formalize your company's commitment to philanthropy (or for any other reason); this trip appealed to us, however, because we like to experience different cultures, and this Native American ritual was something we wanted to explore. Like our white-water rafting trip on Idaho's Salmon River back in 1988, this vision quest was successful because we know what works for *us*: going away, going off-site, finding a place for introspection. If you want to get away from your daily responsibilities to think about the bigger picture of your life, your relationships, and your business, you need to find what works best for *you*.

Find Good People to Guide You

Alan: Once we decided to set up a charitable foundation, we knew we would need some guidance. On the flight home from Arizona, we made a list of people we admired who promoted peace, education, and environmental stewardship all around the world. Today, we would call them social entrepreneurs—people who have the vision and courage to take risks and adopt innovative approaches to creating social change. We wanted people like that to serve on our board of advisors.

We aimed high: at the top of our list was Sir Edmund Hillary. We had long admired the spirit of adventure that took him to the top of Mount Everest; he is the first person known to have reached its summit, back in 1953. We shared his love of the Himalayas and the Sherpa people. In our eyes, Sir Edmund Hillary was a powerful role model for leadership and philanthropy. *Time* magazine even named him one of the 100 most influential people of the twentieth century. Over the years, hundreds of people have climbed Everest, and many succeeded in reaching the top, but only Hillary made it part of his life's work to help the local Sherpa people overcome their own challenges. His relentless fund-raising and keen oversight of the Himalayan Trust, which he founded, have been an inspiration to us all at Grand Circle Foundation (GCF). We were thrilled when he agreed to join our board of honorary advisors.

We also invited people from several charities and organizations that worked in places where we had lots of travelers. We had a connection to these destinations, and we wanted to give back to them. Finally, we asked other friends and fellow travelers to be part of our advisory board.

Once we had our core group of advisors, we began formalizing our arrangement. We initiated the legal steps to create a charitable foundation tied to Grand Circle Travel. We committed to contributing at least $10 per traveler, per trip, toward the various projects the foundation would support.

Finally, we knew we needed to find someone specifically to lead the foundation; this wasn't going to be a responsibility we could fold into someone else's job. Our first director of the foundation was my cousin, Roland Barth, an educator and author. Later, we hired Martha Prybylo, who eventually became our executive vice president heading up our People & Culture department. And after that, Maury Peterson, a former Peace Corps volunteer, came on board; Maury is passionate about social change and is someone who can get things done.

Back then, we had a vision, a mission, and a great group of advisors. It was time to get to work.

Start Slow: Don't Take on More Than You Can Handle

Harriet: We began slowly, giving small amounts to organizations around the world as a way to say thank you to the countries that had so warmly welcomed our travelers. We didn't have a real plan at all—just to help. We had no clear concept of what we were doing. In fact, in the first year, we gave only $16,500.

But we soon picked up the pace. After we acquired Overseas Adventure Travel in 1993, we extended our reach because there was so much poverty and need in the destinations that OAT tours visited, such as East Africa, Tibet, Turkey, Morocco, and Patagonia. We set up a teacher-training program in Nepal. We raised a totem pole in Alaska. We saved rhinos in Zimbabwe. We funded a clinic in the rain forest of Peru. We helped rebuild war-torn Dubrovnik, the beautiful city on the Adriatic that so many of our travelers had discovered on our trips in Yugoslavia. We sent medicine to Rwanda. We provided earthquake relief to Turkey. We supported a cancer hospice in Spain. We bought ceiling fans for a school in Fiji. We gave computers to Buddhist monks in Thailand.

Everything was piecemeal: we simply gave to organizations where we had travelers. We weren't too focused or specific in the beginning (over the years, we have learned to be more focused, specific, and intentional about our giving). Back then, we were just happy to help. And everything felt good.

Pretty soon, we were overseeing dozens of projects all over the world. We found that giving money away wasn't as easy as we had imagined. We had big hearts and were giving money to everyone . . . but in the early years of our foundation, we lacked clarity around how the money was to be used. And we had no measurements to ensure we were making a difference. We were robbed more than a few times. At least once, one of our own regional offices skimmed funds from the foundation. And our money didn't always go where it would be most useful.

Willy Chambulo opened our eyes to the problem with one of his typically casual but profound remarks. Willy is the entrepreneur owner of Kibo Guides in Tanzania. He is a friend and business partner, and he now serves as an honorary board member of the foundation. When we were considering funding a science lab at a school in his country, he warned us, "This is Africa; people are hungry." We knew instantly what he meant. If you throw money around where people are hungry, much of it will be peeled off as it passes through different hands—unfortunately, not by the people who are literally hungry. In desperately poor countries, some well-fed folks often feel they are entitled to a piece of every economic activity.

Willy taught us another valuable lesson: understand how far your American dollars will go abroad. In many developing countries, a dollar has colossal purchasing power. That means more can be done with less money than you might think. It also means that unscrupulous people can pad estimates and fiddle the books so that they can embezzle from naive Americans—and they seldom get caught.

If we knew then what we know now, we would not have given money so freely. We would have had some checks and balances in place to ensure that what we donated actually made its way to the intended projects. We would have made sure people were accountable. But as the old saying goes, hindsight is 20/20. So we learned as we got more involved.

If You Want Philanthropy to Work, You Need a Plan

Alan: Over time, our efforts became more concerted. We saw the value in working with established charities and nonprofit organizations that had knowledge of local problems and could offer proven solutions. From this insight came our early partnership with the World Monuments Fund, which helps preserve and protect ancient and historic sites around the world. Another early partnership was with AmeriCares, the disaster relief organization that provides immediate emergency medical care internationally. We also partnered with the Monteverde Conservation League and with Sir Edmund Hillary's aforementioned Himalayan Trust. All these organizations taught us a ton. We chose these organizations for a variety of reasons. We had personal contacts there. Each of these organizations had a strong history of legitimate business practices. And they were located in areas where we did business, so that was another direct connection to us.

We also made another big change. We involved our travelers directly in our philanthropy. We winnowed our list of projects to those that fell literally along the path of our trip itineraries. Then we designed our trips so that the travelers could visit our projects and see the foundation's work in action. Soon, our travelers were visiting the tombs we helped restore in Egypt's Valley of the Kings and discussing the Holocaust with the director of the State Museum of Auschwitz-Birkenau, to which we have given $400,000 over the years. By 1995, the same model was playing out in schoolrooms, craft workshops, historic churches, child-care centers, and game reserves all over the world. Our philanthropy was no longer conducted on the sidelines; it was now directly in the path of our travelers. It had become a strategic part of the business.

It's a win-win for the people we help and for Grand Circle. We believe that philanthropy should be a part of *everyone's* business; it *is* strategic. If you are going to give anyway, why not give to something that benefits your customers, your employees, and your company?

We knew our philanthropy would benefit local people and interest our travelers. But we didn't anticipate how often the work of Grand Circle Foundation would actually change our travelers' lives—and the places they visit. One of our travelers, Susan Rickert of San Francisco, was so moved by the appalling condition of a school she visited in Tanzania on one of our trips that she spent the next several years raising $100,000 for five schools and finding sponsors for 17 student scholarships. Two more travelers raised enough money to donate 40 bicycles to schoolchildren in the Tanzanian village of Karatu. The bicycles get the children to school and help their families transport such necessities as water and groceries.

A couple gave $5,000 to a school in Don Chum, Thailand, which will help fund food, clothing, bicycles for transportation, and the salary for an English teacher. One traveler, who had gone to Argentina with OAT a few years earlier, read in our foundation newsletter that a school there had been damaged by flooding, and he donated $2,000 to the school—in lieu of Christmas gifts to his family. Another couple was so moved by the needs of Vietnam's Minh Tu Orphanage, where an erratic electricity supply often left the children in the cold, that they donated $3,000 toward the purchase of a generator.

Another traveler was so impressed by the students and faculty at Costa Rica's Sonafluca School that he offered to match all donations made to the school within a two-month period, up to $5,000. Then a different traveler on the same trip offered to donate $8,000, which prompted the first traveler to increase his match to $8,000. That's an enormous amount of money for a rural school like Sonafluca, which used that $16,000 to replace classroom windows and doors, renovate the bathrooms, furnish the cafeteria, construct a covered walkway to keep the children dry during the rainy season, and do much more.

This kind of deep involvement with people who were once strangers happens to our travelers all the time. These are only a few examples of literally thousands. More than 14,000 of our customers gave $1.1 million in 2011 alone. Our travelers say we inspire them, but they inspire us, too!

What could *your* company do if you found a way to partner with your customers to help some part of the world less fortunate?

Stay Focused on How You Want to Give Back

Alan: Our honorary advisors were totally committed to helping people, but they also saw themselves as advocates for their own corners of the world. In the early years of the foundation, this kind of local interest posed no problem because our resources were limited and their enthusiasm made our efforts much more effective. But as the foundation grew, we wanted to branch out to places that our advisors did not represent. We wanted to give back in Thailand, East Africa, China, Mexico, and Peru, for example. Our advisors continued to lobby for their own interests, though, and we had more and more trouble getting new projects off the ground.

At the same time, we stopped seeking their counsel and didn't listen to them at a high level, where we really needed the benefit of their experience. Our advisors were mostly interested in advocating for the places and organizations they felt we should be supporting. That was helpful, but we also needed hands-on leadership and direct involvement with the charities we were helping. We learned that *leadership* is much more important to philanthropy than mere *advocacy*. When you get a strong leader and a passionate advocate in the same person, that's when philanthropy really catches fire.

We had plenty of problems in those early days. The turning point came when Harriet got personally involved in a school project in Costa Rica in 2005. Harriet had served as the chairperson of Grand Circle Foundation from the beginning. She had often gone into the field to evaluate projects and approve donations. Harriet had first become interested in the problems of schools in other countries in 2001 when we visited Sinya, Tanzania (near Kenya's Amboseli National Park and Mount Kilimanjaro). We had stopped by the local school, as we usually do, and Harriet was dismayed. The schoolroom was tiny and filled with hornets. Students shared space with stored food and other provisions. The books were moldy.

After observing class for a few hours, she asked herself, "What chance do these kids have?"

We knew we had to do something. But we wondered how much difference we could make alone. Luckily, we knew Willy Chambulo (whom we have mentioned before), a social entrepreneur in Tanzania who had built his company Kibo Guides from scratch. Willy also loved to build physical things. We had been partnering with him for years as a ground operator; now we decided to take our relationship to a new level. We made Willy a proposition: we would help capitalize him so he could build wilderness camps and help us fix up the school in Sinya, and he would pay us back with reduced rates when our travelers stayed at his camps. We saw this as a way to buy direct in Tanzania and simultaneously help the children of Sinya. We would work the business side through both our Overseas Adventure Travel and Grand Circle Travel divisions. And we would work the social side through Grand Circle Foundation. Willy did a magnificent job building the lodges and school, and we've had a loyal partner and true friend ever since.

Then, in 2005, Harriet became aware of a new project involving six schools in Costa Rica. This project held special interest for her because she had trained as a special education teacher at Kent State University and had worked with special-needs children for many years in inner-city schools. Here was an opportunity for her to provide direct leadership in an ambitious new partnership.

Harriet had the support of Dr. Rodrigo Carazo, the former president of Costa Rica, who was one of our board members. She also had logistical help from local Grand Circle staff. Their contributions to the project were important, but Harriet soon saw the need for something else—*local community involvement.* For example, she met with parents of children at the San Francisco School, located just outside of Costa Rica's capital city of San José. Those parents asked the foundation to donate computers for the school. But there was a problem: the classroom was open to the elements, and it had neither electricity nor air-conditioning. That was a huge obstacle at a time when computers overheated easily.

In our various charitable efforts over the years, we had learned we can't think only with our hearts. We also need to be realistic about what we can accomplish with finite resources. So Harriet sat down with the parents and teachers and asked them to commit their own talents and energy to building a reasonable computer lab. And they did. Many of these parents were already very committed: one father, for example, had been driving his daughter from the village into San José to learn about computers—three hours for the drive and lesson each time. One parent offered to pay for the electricity. Some of the parents built the walls. Others ran the electrical wiring. Then they put on a series of fund-raisers to pay for air-conditioning.

By the time the computers arrived, the community had already invested a lot of its own time and money. The community had ownership of the project, and all the parents made sure their children treated the equipment with respect.

Harriet had pioneered a new model of *reciprocal responsibility* for Grand Circle Foundation. Such a model allowed our partners the freedom and independence to meet their own goals. And we had found a formula for global philanthropy that worked. Strong local leadership and a partnership based on shared responsibility made all the difference.

Harriet's work in Costa Rica laid the groundwork for our World Classroom initiative, which launched in 2005 with the goal of fostering children's education and engaging the communities in which they live. Today, the World Classroom initiative supports nearly 100 schools in 60 villages in 30 countries.

This initiative is a program that resonates particularly strongly with our travelers who visit these schools. They care deeply about educating children and giving them opportunities for advancement in the global village. Many of our customers have gotten involved directly over the years with the schools they have visited on our trips. We're grateful for the help and support they have contributed to educating children everywhere in the world.

Develop Your Own Rules of Philanthropy

Alan: Many lessons came out of the Costa Rican school project, and as we took the World Classroom initiative worldwide, we found we could apply the lessons almost everywhere we went. Now we call them "Harriet's Rules of Philanthropy." They consist of five guidelines for global giving that reflect our experience with many projects that worked. (And some that didn't!)

1. Get personally involved.

2. Work with strong local leaders on common goals.

3. Foster respect, reciprocity, participation, and independence.

4. Know the value of a dollar.

5. Engage social entrepreneurs to magnify giving.

We believe these guidelines will work for any organization that wants to make a difference in the world. Let's take a closer look at each.

Rule 1. Get Personally Involved

Harriet: Global philanthropy is more than long-distance charity. Money is important, but close involvement with projects—especially overseas projects—stretches your dollars further. It ensures that your donations are really meeting local needs and that the recipients know their project is truly important to you. And it lets the local people know there will be oversight. Obviously, you can't personally supervise every project. So you need to get commitment from regional or local associates before you recommend a new endeavor in their areas. Then they will become your eyes and ears on the ground. Since my first experience in Costa Rica, this is what has worked for us, and we believe it can work for anyone with regional offices—whether within the United States or around the world.

In fact, this lesson literally came home to us when we tried to help with a project much closer to home that failed. We have a home in Kensington, New Hampshire, that has been in the Lewis family for four generations, and so we have deep roots in the community. Kensington is a small township with only 1,800 people in a 12-square-mile area. We love it, and we wanted to do something for the community. So in 2004, we began discussions with the town to build a nice park on some land we owned there. With all our experience with Grand Circle Foundation, you would think this little philanthropic project would go smoothly, right?

Alan: But it didn't. I broke a few of our rules of philanthropy, and they came back to bite us. For one thing, I was busy running the company and our 30-plus offices around the world, and so I didn't have time—or I didn't make the time—to get personally involved. Instead, I appointed a liaison to work with the town. I knew the guy, but somehow he managed to antagonize everyone.

Another mistake we made was to throw around a lot of ideas for the park without really listening to town officials or asking what the residents themselves wanted or needed. There were too many chiefs, not enough trust, and no shared vision for the project. So at that point, we just walked away.

A couple of years later, one of the early project supporters approached us, hoping to get the park back off the ground. This time, things were different. The selectmen (selectmen are typically a New England phenomenon—similar to members of a town council) formed a committee of town officers, youth and recreation staff, and representatives from our Grand Circle Leadership Center, which is also located in Kensington. We listened harder this time. And we got a new point man, a Kensington resident, who not only was a strong project leader but also got along very well with everyone. Everyone involved was a townie, and before long, the project got neighbors out of their homes. People were united in the excitement of getting the job done.

One day, Harriet said to me, "This is just like Costa Rica. The project didn't really get started until we started listening, found a good leader, and made sure everyone in town got involved. Good philanthropy *does* take a village."

Sawyer Park opened in 2008 (it's named for my mother). It's even grander than we first imagined. It has three lighted baseball diamonds, an all-purpose field, a basketball court that converts to an ice-skating rink in winter, a playground, and a skateboard park. It also has bathrooms, a full-kitchen concession stand, and walking paths that connect with the town trail system. The entire park is organic—no pesticides or herbicides allowed. Four fertilizer companies each adopted a field, and residents help with turf care.

We donated the land and the funds to build Sawyer Park, and we make annual contributions toward its maintenance. It's been worth all the early trouble to watch the community come together under the lights for hot dogs and a Little League game.

Harriet: We learned a lot, too. You need to listen before you can work together, and you need a strong leader to keep projects moving along. This is one of the most meaningful philanthropic donations we've ever made, and we've made a lot. Who knew it would take a world of giving to know how to bring charity back home? It could have been a success so much sooner if only we had paid attention to Rule 1: *Get personally involved.*

Rule 2. Work with Strong Local Leaders on Common Goals

Harriet: We broke this rule also, on the first go-round with the Sawyer Park project in Kensington. Fortunately, we remembered it on our second effort to create the park. That time around, we partnered with local leaders in Kensington to ensure they got the facilities that would serve them best.

Strong local leaders are excellent community organizers, and we always try to get them on board at the very start of a project. That way, the excitement that outside support creates in a community can be given focus and direction. In many of our village projects around the world, those leaders are school principals, small business owners, or elected officials. In other words, they're people who know the local political landscape and can build community participation, which is essential to the long-term success of our projects.

But local leaders cannot work effectively in isolation. So it's important to work as closely with them as they do with the community. We learned this from Sir Edmund Hillary. He had a common goal for every project he undertook for the Sherpa people in Nepal: *sustainability*. He felt strongly that once a project is complete, it should belong to the community. That is only possible if the project is designed in a way that the community can sustain it without ongoing financial support. It's like the old Chinese proverb:

> **Give a man a fish and you feed him for a day.**
> **Teach a man to fish and you feed him for a lifetime.**

We have always taken Sir Edmund's philosophy to heart in the villages we support through Grand Circle Foundation.

We formalized this commitment in 2009 when we launched our Invest in a Village initiative. We have always been deeply involved in villages around the world. For example, our Day in the Life program takes our travelers into homes, markets, fields, and craft workshops around the world. We've had this program since the earliest days of the company, even before we established Grand Circle Foundation in 1992. In 2005, the World Classroom initiative extended our local reach to village schools.

Our Invest in a Village initiative goes yet another step. Here, we partner with leaders in villages where we have already developed strong relationships to support the growth of entrepreneurial opportunities based on local community needs. The model for this program is our microfarm at the San Francisco School in northern Costa Rica. This is a three-phase project we began funding from the very first seed in 2006. The farm provides hands-on learning opportunities for students. It also supports the entire local community with high-quality nutrition and revenue from the sale of surplus produce, which goes to support the school. It also gives children from agricultural families a reason to stay in the village after graduation, rather than leaving to find work in San José. That was a pattern that greatly concerned their parents, who wanted them to stay.

In another Invest in a Village project, in 2010, the foundation set up a women's sewing workshop in Bearat, Egypt. That project raised money to buy 37 sewing and embroidery machines so that the village women could create traditional Egyptian *galabeya* dresses, school uniforms, and table linens to sell. This project built on a tradition of hand sewing in the area. But the new project increased both output and revenues, and it created jobs for at least 35 women who were previously unemployed.

In these ways, we hope to create sustainable change through local entrepreneurship.

Rule 3. Foster Respect, Reciprocity, Participation, and Independence

Alan: Early on, we discovered that many Third World communities are so poor, they will take any kind of donation, even if what's being donated isn't the thing they need most. If a charity is offering computers, people will take them even if what they really need is clean drinking water. Sir Edmund Hillary reinforced this lesson: he told us that it is always best to begin a philanthropic enterprise with a face-to-face discussion. "Ask local people what they need," he said. "Don't tell them what you *think* they need." Not only is this more respectful of local communities, but it is also a better use of your money.

In our work, once we decide on a project, we work hard to develop the kind of reciprocal responsibility that Harriet pioneered in Costa Rica. This means asking for community participation. We ask the local people to help plan the project. We ask them to volunteer tools or labor. Or we might ask them to raise additional funds. This kind of community participation mirrors our own personal involvement. It also creates ownership in the project from both sides. In this way, our projects become partnerships among equals, and the projects have a better chance at long-term sustainability.

Rule 4. Know the Value of a Dollar

Alan: Knowing the value of the U.S. dollar means understanding its purchasing power in every region you give to. It also means not overcommitting funds. It means keeping tight budgets. It means demanding accountability. And it means safeguarding the money channel—in other words, making sure the money you're giving gets into your recipients' hands as directly as possible. Eliminate the intermediaries. In many parts of the world, that also means eliminating official government agencies. Lax financial practices just invite graft. Keep a tight fist and a watchful eye, and you'll be able to give more money to more people who need it.

Rule 5. Engage Social Entrepreneurs to Magnify Giving

Alan: Whenever possible, we partner with social entrepreneurs who understand local issues. They also have experience delivering money and services in the regions where we give money. Besides being principled people and gutsy leaders, these entrepreneurs can give us some hands-on professional help. They supervise distributions, they meet schedules, and they audit finances. Some of our partners are businesspeople, like Willy Chambulo. As noted earlier, Willy was the one who warned us that hunger would divert our investment in Africa if we were not vigilant. Others are village elders, local community workers, and even our own travelers and guides. We believe entrepreneurial leadership can come from anywhere. And we are constantly on the lookout for it.

All these rules were learned in the school of hard knocks, and we made plenty of mistakes along the way. We still make mistakes, but there are fewer of them, and we catch them earlier now that we know what they look like. You will also likely make mistakes in your early philanthropic endeavors. But we hope you will learn from our mistakes as much as possible so that you can truly help the people and organizations that will benefit most.

Encourage Philanthropy in Your Own Backyard

Harriet: Our foundation has one single mission—to help change people's lives. But it plays out in two spheres: in the world we travel and in the world in which we live and work. The first sphere is global. So is the second, for we are an international company with 34 offices worldwide.

Still, the philanthropy we engage in at home is often different from the work we do abroad. Why? Because it happens in our own backyard, among people and in places we might see every day. It is the realm of community service.

In Boston, the tradition of our community service actually predates the start of Grand Circle Foundation. It goes back to 1985, the year we bought the company. From the very beginning, we encouraged our associates to "give back" by putting their time, money, talents, and energy into local causes they believed in. We wanted them to experience the pride and connection that community service work brings. We also wanted to afford them leadership opportunities as team captains, fund-raisers, and project organizers.

Since giving back is part of our business strategy, we make it easy for associates to serve. We develop a yearlong calendar of events from which they can choose, and we help set up the events. For example, in 2010, the foundation sponsored 21 different events for the Boston office—and a record 97 percent of our Boston associates participated. In addition, 26 of our 34 worldwide offices organized community service events of their own. Many of those service events benefited schools and villages supported by Grand Circle Foundation.

We also publicly recognize our associates' efforts in our newsletters, at company meetings, and at annual awards ceremonies. In the late 1980s, we enshrined community service in our mission, as part of our commitment to social responsibility. In 1993, we created a community service team (CST) to help organize associates' service work. The CST consists of a core group of 20 to 30 associates who plan volunteer events for Grand Circle. There's also a dedicated staff person who works for Grand Circle Foundation, and she helps drive the events we do each year. We're very pleased with how the program has grown from about 8 to 10 events per year to about 25 per year. The GCF person usually gets one or more associates to chair each service event, so that associates at every level get a chance to lead an event and hone their leadership skills.

In 1994, we gave $25,000 in seed money for our Associates' Fund, which makes its own grants to Boston-area nonprofit organizations. This fund is coordinated by a GCF staff person and a group of associates who volunteer to solicit, review, and vote on grants from local nonprofits. The volunteer team meets and decides on grants twice a year. Alan and I provide the funding, but the associate grants are decided (and awarded) by the associate volunteer team. In some years, we have given as much as $100,000 to support extraordinary efforts.

We are amazed by how enthusiastically our Boston associates have taken up the challenge. Over the years, we have joined them as they walked for AIDS, biked for cancer, swam for clean harbors, and ran marathons for more causes than we can count. We've given blood, sold pies, cooked meals, collected toys, painted houses, planted flowers, read to toddlers, sat with elders, and helped teenagers with their homework.

In 2006, one of our associates started a new program called Up & Out. This program built on our longtime support of the Boston Family Shelter. Up & Out helps homeless families move into apartments by helping them acquire furniture, bedding, kitchen appliances, and even cleaning products. In 2010, the program relocated its twenty-fifth family. In 2011, we shifted the focus of our Up & Out program to homeless military veterans after an event we did for the New England Center for Homeless Veterans generated requests from our associates to do more to serve our local veterans.

Within the company, community service has spread all over the world, for as Grand Circle grew, so did our "backyard." We opened our first regional office in Germany in 1997; today, we have 34 regional offices from Cairo to Killarney. Community service looks different in each region, as our associates respond to different local needs. In Thailand, for example, associates once built a chicken coop, while in Mexico, a group of associates drove a mule team into the Copper Canyon to bring clothing to a remote Tarahumara village. Associates have baked moon cakes for elders in Hong Kong, renovated a classroom in Vietnam, and purchased musical instruments for children in Italy. It seems each office is developing its own tradition of giving. That's what we want them to do. It's the best way to ensure that the service is both heartfelt and helpful.

Over the years, Grand Circle associates have volunteered more than 40,000 hours to community service projects worldwide. And they've donated more than $1 million of their own money to local causes. Participation is voluntary, but enthusiasm is contagious.

Today, more than 90 percent of our Boston and overseas associates do some community service work every year, including 97 percent in 2010.

Here's just one recent example. When a devastating earthquake hit Haiti in January 2010, we sent an e-mail appeal to our associates and travelers asking for help raising money fast. Grand Circle Foundation offered to match any donations received within two days—up to $150,000. Within 48 hours, 2,500 travelers and many associates had responded with contributions that totaled nearly $400,000, including the foundation's matching funds. Follow-on donations exceeded $100,000 from an additional 1,500 travelers. That's 4,000 travelers who responded to our appeal right away, helping us to raise $541,000. What a great bunch of travelers and associates to partner with us so enthusiastically! See what we can do when we work together?

Personally, Alan and I continue to donate to the causes we believe in. Over the past 25 years, we have given close to $15 million to Boston's West End House Boys & Girls Club, Artists for Humanity, Boston Children's Museum, the City on a Hill Charter School, Summer Search, Big Sister, Thompson Island Outward Bound Education Center, AIDS Action Committee, and Greater Boston Food Bank. It was hard to winnow down so many great organizations to a manageable group that we could make a difference to, but we finally did it.

Nurture Social Entrepreneurs

Harriet: In 1999, Alan gathered many of these long-term community partners to form the foundation's Community Advisory Group (CAG). At first, the group met to advise Grand Circle Foundation staff on its Boston grants. In recent years, however, it has become a kind of incubator for social entrepreneurship, a place where nonprofit leaders can meet informally to share war stories and best practices. We also attend the meetings, where we challenge the nonprofit leaders to get clear on their vision and mission and to become more focused on their leadership development. It is an innovative and collaborative model for philanthropy, one that seeks to pool experience to raise individual and collective performance.

For example, one of our community partners, Greg Zaff, wanted to use his love of the game of squash to help young people from the inner city become better students and responsible citizens. So he founded SquashBusters, a nonprofit group that would do just that. The squash was the easy part (Greg is a pro), but the schoolwork and citizenship proved harder. As a member of CAG, SquashBusters received help with this on both fronts from fellow CAG members. One organization in the group set up the Squash-Busters teaching program. Four others offered community service projects for the kids. Greg also brainstormed with fellow group members to improve SquashBusters' grant-writing and fundraising efforts. "No other funder in the city does this kind of collaborative work," Greg says. "There is real power in the informal learning, encouragement, and inspiration we get working with other CAG members. It has been so energizing; it has given us so much perspective, and we've met really tremendous people. This is a new kind of philanthropy—open, visionary, and empowering—and it has really made our work better."

$91 Million Goes a Long Way around the World

Alan: Since 1992, Grand Circle Foundation has donated or pledged more than $91 million to more than 300 projects worldwide, some of which are listed in Appendix A. Today, we support projects in more than 40 countries, including 100 schools in 60 villages in 30 countries. And our company has been recognized for our efforts (listed in Appendix B). We feel really good about it.

We all want to be part of something bigger than ourselves. It's human nature. That's why professional sports teams draw ardent fans and why designer clothes are so popular. Companies that can turn their customers into fans also do extraordinarily well. Just look at Apple electronics, BMW, Ben & Jerry's, Google—and Grand Circle. Because we have integrated philanthropy into our business model, taking travelers to visit our projects around the world, we have made strong emotional connections with our travelers. We know that's one of the reasons they travel with us again and again. We know because they tell us so in their trip evaluations—and we pay very close attention to what our customers tell us in these evaluations. Our travelers also support our philanthropy through unsolicited donations to Grand Circle Foundation—more than $1 million to date.

Another benefit of our philanthropy is that it gets us help from local people when problems arise in faraway places. Grand Circle is considered a gracious visitor all over the world because we support communities in the countries where we travel. Our regional associates appreciate the recognition they receive in their countries for directing our donations. And they are proud to work for an American company that encourages them to become involved in local philanthropy. All of this goodwill pays dividends when we need effort or support above and beyond the norm. We don't engage in philanthropy to elicit favors, or to build morale, or even to attract travelers. But we've seen the results.

Here's our advice: *if you want to build a great company, make philanthropy as important to your business as marketing and finance.* If you don't feel the impulse for philanthropy in your own heart, give the responsibility to someone in your organization that does. You will find that you'll do very well by doing good.

Delivering Unsurpassed Value

Focus on what you do best,
not on what you can do.

lan: As we grew Grand Circle, we knew that in
order to be a great company, we needed to con-
trol two things: the *quality* of our products (which we refer to as
"excellence") and their *value* to the customer. We learned that we
needed to keep a tight focus on what we were offering, concentrat-
ing on what we do best and not on all the things we *could* do. We
consider that our relentless attention to excellence and to unsurpassed value is one of our Extreme Competitive Advantages—one
of our six major strengths that keep us three to five years ahead of
our competition.

Winnow and Improve Your Product Line

Alan: When we bought Grand Circle Travel in 1985, it had only
16,000 customers (compared with more than 115,000 today), yet it
offered about 500 different trips. Even worse, many of these trips
were special-order itineraries that might never be offered again. The
trips we inherited from Colonial Penn were a crazy jumble of sightseeing excursions, countryside vacations, tours of foreign capitals,

extended stays in apartment hotels, and mad-dash "passport-stampers." Clearly, we needed to focus on which of those trips were most profitable, and we needed to ditch the rest.

Much of our work in the first two years that we owned Grand Circle was to get those trips under control. Our first action was to get rid of the worst-performing products. We looked at the profits and losses for all our trips and immediately saw that more than 300 trips were unprofitable or of poor quality. We cut them all. We were merciless because we had to be, and the instinct to cut our losses and put our money on the sure thing has stayed with us over the years. Once the poor performers were out of the way, we were able to focus the entire organization on the trips that were popular and profitable and of high quality.

At the same time, we knew we had to increase the value of our products. So we stopped using travel agents to market them. Instead, we marketed directly to the AARP list, saving our customers the cost of the agents' commissions.

Next, we slowed down the pacing of the trips that remained. Nobody likes getting predawn wake-up calls or being marched from one famous site to another. Well, some people do—some tourists are competitive sightseers who want to punch their ticket for as many sights as possible, even if the bus stops only momentarily. But we wanted to attract a different type of customer, a different type of traveler. We wanted our travelers to have time to enjoy the people and cultures in our destinations, maybe explore a back-alley bazaar or spend time with students at a rural school. That couldn't happen if they were cooped up in a big bus or if the tour guide held a stopwatch every time they got off. In fact, *pacing* became one of our four "Product Pillars," described later in this chapter.

We scrambled to make these changes quickly, partly because we needed to cut our losses and partly because we were cocky. We believed we could radically reduce our trip offerings but still keep our customers. We thought our new lineup of well-paced, engaging, value-priced trips would appeal to our customers, and we were right. By the middle of 1987, two years after we acquired Grand Circle, the company stopped hemorrhaging $2 million a year, as it had been doing, and instead began to turn a profit. We were ecstatic but cautious: we knew we needed to focus our product line and customer base even further.

Excellent Products Are Key to a Great Company

Harriet: Earlier in the book, we described how we focused our vision for our new company by going on an off-site trip, white-water rafting on Idaho's Salmon River in 1988. The discussions we had during that trip defined our product strategy, but now we needed to take action to make our new strategy a reality. We had whittled down our product line and stopped the bleeding. But now we needed to take a harder look at our trip offerings, this time with an eye to weeding out the "cheap and nasty" vacations and identifying the trips that could offer really meaningful experiences.

That meant more cuts. And this time, it was much harder. Our associates had been running the trips for more than three years, and they had become invested in them. Over the years, we have learned that people get attached to what they know. They'd rather fix what's broken than move on to something new. Someone always seems to step forward to defend a floundering product, giving reason after reason why it's a good idea to give the product a little more time. But that's not the way to build a strong company.

Focus on Your Most Profitable Products for Your Best Customers

Alan: Profitability depends on delivering unbeatable value, and value depends on constantly improving your best products. In our business, if a trip is of poor quality or explores a destination people are not interested in, we cut it; if we don't, it becomes a distraction. The same is true in any business—whether you own a bakery and *you* love peanut butter cupcakes but your customers aren't buying them, or if you're a dentist who wants to specialize in cosmetic reconstruction but your best customers are families with kids with cavities. You need to focus on your best, most profitable products that will appeal to your best repeat customers and keep working to make them even better.

So that's exactly what we did back then and have been doing ever since. We cut and we built. We eliminated the last of the Colonial Penn products that didn't suit our new product strategy. We consolidated the best features of some legacy programs like our Extended Vacations. These vacations took travelers on two- or three-week tours that allowed them to stay a week or more in each destination; on some itineraries, travelers could purchase one or more additional weeks for less than $100 a week, and on Spain's Costa del Sol they could stay half a year.

Extended Vacations met our criteria for slower pacing and discovery, and we rolled them out to new destinations: Interlaken, Lucerne, Gstaad, Guadalajara, the Amalfi Coast, London, Paris—even Yugoslavia, a country Americans seldom visited in those days. When the Yugoslavia trip became a bestseller, we knew we were on the right track. The vision and mission that caused such a fight on the Salmon River were fast delivering results.

While we made the right decision to cut more than 300 trips and expand our Extended Vacations program, we also made quite a few mistakes along the way. It's possible to cut too deeply, which we did in Russia. It was a big mistake (as we'll describe in Chapter 6), because it cost us a lot of money to get back into the country and made it difficult to achieve the strong value we might have had if we had never left.

Differentiating Our Products in Four Key Ways

Harriet: Over the course of about four years (and from our discussions during several off-site meetings), we established a new vision for our company: we would offer meaningful cultural encounters at unsurpassed value that could actually change people's lives, and we would do this by building all our trips around what we call our four Product Pillars: value, pacing, choice, and discovery.

1. Value

Harriet: Our products offer not just good value, but *unsurpassed value*. And we're not talking only about price. Instead, we look at value as the combination of price and the *experience* you get for that price. The easiest thing in the world is to cut back on trip features until you have the rock-bottom price. But what's the point of traveling at the bottom? You'd get a better experience watching a travel special on TV.

Other business leaders have realized the same thing. Gordon Bethune is one; he's the former CEO of Continental Airlines who turned the company around from its worst days to become profitable and successful again. Gordon once quipped that "you can make a pizza so cheap that nobody wants to eat it." We agree . . . and we don't want to offer products that nobody wants—which, in our business, means trips that no one wants to take.

Instead, we strive to keep our price as low as possible while still maintaining excellence and offering high-quality features. For example, a home-hosted meal with a local family costs less than a restaurant meal for the group, but it gives our travelers a much more authentic cultural experience. On a per diem basis, we offer the best value in the industry, without exception. For example, on our bestselling river cruise from Amsterdam to Vienna, our travelers pay at least $100 less per day than they would with our strongest competitor, saving between $1,000 and $2,000 per trip.

2. Pacing

Harriet: Good pacing makes for an enjoyable trip, and that goes double for older travelers (the average age is 72 for Grand Circle travelers and 65 for Overseas Adventure Travelers). We pay special attention to the first 48 hours, when travelers are tired from their overseas flight. We limit early wake-up calls whenever we can. We train our trip guides to stroll at a leisurely pace. When we travel by road to a featured site, we try to find a route and time that avoids the worst traffic. We don't jam everything possible into an itinerary, because memorable trips allow travelers time to savor small events like sipping coffee at a sidewalk café. As a rule, we spend at least two or three days in each location on land trips—on some GCT trips, six to eight days—enough time to settle in and feel comfortable.

3. Choice

Harriet: We know Americans relish their freedom and independence, and so we build a lot of choice into every trip. We offer escorted land trips, river cruises, small-ship coastal cruises, extended vacations, and small-group adventure trips in 100 countries around the world. Each trip can be customized with a choice of optional excursions. Travelers can also add days at the beginning or at the end of a trip so they can do as they like on their own. We build in plenty of free time for travelers to follow their guides' advice on good restaurants and sights to see. Open seating on our river cruises lets travelers explore different vantage points on the ship and sit with whomever they please. We also encourage our guides to offer choices to travelers on the fly, so that if unexpected opportunities arise, they can decide whether to follow them. We have always found that the most successful kind of leadership empowers others to make their own decisions.

4. Discovery

Harriet: Discovery is the most important pillar, the one that helps us create unforgettable experiences. When Alan and I travel together, we like to interact with local people and discover the local culture firsthand. I like to poke around markets and talk with teachers, whereas Alan likes to take long walks to find places tourists don't usually see. We assume our travelers want to do these things, too. So over the years, we've created a series of signature discovery events for our travelers. We visit schools. We offer home-hosted meals. We go on excursions to native markets and bazaars. We have exclusive tours of some of our Grand Circle Foundation projects. We cook, do crafts, and have discussions with experts on hot-button local political and economic issues. These kinds of programs can be difficult to set up, but they pay off in customer satisfaction. We ask all our travelers to fill out a survey at the end of their trip, and when discovery is high on a trip, we get high scores on those evaluations. When we fail, we see low scores, and we know we have work to do.

Getting to Excellence by Providing Unforgettable Experiences

Harriet: Fortunately, over the years, we've improved our product line so much that our travelers today tell us we have excellent trips: in 2010, for example, 81 percent of our travelers rated their trip as "excellent." We use three tools to achieve excellence:

1. We design all our trips around our four Product Pillars.

2. We evaluate our trips using the best quality-measurement system in the industry.

3. We hire and train the greatest guides in the world.

We take a three-pronged approach, and each component supports the others. The Product Pillars give discipline to our trip design and delivery. Our quality surveys give us feedback so we can improve the products. And our guides work every day to keep the unforgettable experiences coming.

When we design our trips, our goal is to give travelers experiences they will remember for a lifetime. Unforgettable experiences come in many ways. Sometimes, they come from the cultural encounters or inspiring sights we deliberately build in to our trips. Just as often, they may come from little, unexpected things, too, like having breakfast at dawn on a sun-kissed dune in Morocco. Our travelers' unforgettable experiences often seem to involve food, like grabbing falafel on the streets of Luxor, trying chicken feet for breakfast in China, or drinking snake wine in Vietnam. Unforgettable experiences can also be simple human connections, like buying fruit from a vendor in an open market, or singing along with a bunch of Hungarians to American rock and roll, or trading sign language with Tibetan monks in another bus while caught in traffic, or making friends with fellow travelers. Our travelers tell us about these kinds of experiences all the time.

Here are just a few of the comments we receive from our customers every day:

> *Our Amazon trip was spectacular. Our guide brought some villagers on the boat for a visit—three families, with their children. They were curious about us. They wondered if we were too cold in our air-conditioned cabins and thought it strange that we slept in beds. We showed them the dining room and kitchen and they were amazed with the stove and refrigerator. They had no electricity, so they fished every day and salted some for later. It's good to be home, but we will never forget the lovely, generous people of the Amazon.*
> —MARGARET AND RICHARD S.,
> SIX-TIME TRAVELERS FROM BAKERSFIELD, CALIFORNIA

On an OAT trip to Egypt, I heard music coming from a restaurant. As a retired music teacher, I simply had to stop and listen. I even asked to join the musicians on stage to play the drums. As I jived along with the band, dressed in a galabeya and headdress, I was one of them. It made me aware of the global society we share, even in our own country.

—BILLIE B.,
THREE-TIME TRAVELER FROM AUGUSTA, NEW JERSEY

I have been interested in yoga and Indian philosophy for more than 20 years, first as a yoga practitioner and later as a teacher. Part of my training included studying the origins and philosophy of yoga in India—but all of these lessons came from books and teachers—until I went to India on an OAT trip. I was grateful that our Trip Leader, Sujay Lall, was so knowledgeable and willing to discuss spiritual philosophy with me. He shared readings about different religious beliefs, gave me a Sanskrit lesson, and went out of his way to give me as many opportunities as possible to experience the spiritual side of India. We visited many temples, were introduced to a "holy man," and met with a guru in Varanasi, where we received astrological and spiritual advice. It was a special experience for me.

—CYNTHIA M.,
FOUR-TIME TRAVELER FROM BURKE, VIRGINIA

The great thing about travel is that you don't need to plan every aspect of a trip to make it memorable. The world is a fascinating place. If you do the pacing right and give trip guides permission to stop the bus to pursue unscripted experiences, then serendipity will immeasurably add to the trip, and pretty soon you will have an unforgettable experience. We now know enough to sit back and let it happen. It took us a few years to realize that these "unforgettable experiences" were what set our company apart from other tour operators, but once we did, we focused on that and never looked back.

What's unforgettable about *your* product or service that will become one of your Extreme Competitive Advantages and make *your* business successful?

Going Direct to Improve Product Pricing, Value, and Quality

Alan: Most American travel companies use American tour guides on their overseas trips. We did, too, for the most part, until 1996, when we began thinking about hiring more local guides. We were about 10 years into the business, and we wanted to expand the authentic cultural encounters we were offering our travelers. Could we deliver more unique experiences if our trip guides were born and raised in the destination country? Would their cultural knowledge, national pride, and local connections make enough of a difference to make up for the difficulty and cost of completely changing our system?

As it turns out, we had our answer in our travelers' post-trip evaluations. At the time, we were using both American and local trip guides in some of our more far-flung destinations, such as New Zealand and China. When we compared their quality scores, we found that our travelers preferred the *local* guides. It made a certain amount of sense. Even the most experienced American guide would have trouble delivering the kind of insider view and behind-the-scenes information that a local guide can. But that sample was small: only about 500 people. Would the conclusion apply across all our trips?

It may seem odd, but the scores our travelers gave us about the air-travel portion of our trips supported the case for using local guides. Our air scores were never very good in those days, but they were better when we used a foreign airline. The seats and service weren't very different, but the excitement of going to a far-away place started as soon as the traveler boarded the plane and was greeted by a foreign accent. Travelers liked that immediate connection with a different culture, and they liked the personal touch.

To be honest, we were a little surprised. The image of the Ugly American dominated American tourism in the 1990s, and many travel companies operated on the assumption that American tourists didn't really want to rub elbows with foreigners, that they preferred their cultural encounters filtered through an American guide who could lead them safely through dirty streets and get them back to the hotel in time for cocktails. Most companies co-cooned their travelers, keeping them away from trouble spots and out of uncomfortable conversations because they thought that's what American travelers wanted. We thought our travelers were different—more educated, more curious, and more open-minded. And we were delighted to learn that was true!

Still, switching to all-local guides would be painful. We had close to 300 American guides working for us, and many of them had been with Grand Circle from the beginning. Some were friends. How could we cut them loose? We considered phasing them out, but that could take years, and while we were transition-ing, we wouldn't be able to guarantee our travelers a local guide.

After weeks of mulling the issue over, we made one of the most difficult decisions of our business lives: we let all our American guides go and replaced them with guides native to their destina-tion. A lawsuit from a group of California guides wasn't resolved for five years, and the settlement was costly. It may have been a dif-ficult decision, but the data from our traveler questionnaires after the change showed that it was unquestionably the *right* decision. As we expected, our quality scores soared.

Getting the Best Local Guides to Lead Our Tours

Alan: We didn't just want *local* guides; we wanted the *best* local guides each country had to offer. We had learned a lot about guides over the years. In fact, early in my career with United Travel, I had worked as a guide, taking travelers to Switzerland and the Caribbean. It was maybe not my finest hour in the travel business—I was young, restless, and perhaps not as patient as I might have been—and I certainly didn't get many tips from my customers—but I learned a few things. I learned that leading trips is hard work. I learned that a great guide can make all the difference to a trip and to a company's reputation. And I learned that a fair number of guides are pretty aggressive entrepreneurs in their own right.

Getting the best native guides would be challenging, and so we reviewed compensation in every country. Then we made sure we offered the best total package, including year-round work and high per diem rates. In return, we set challenging goals and rigorously measured our guides' performance.

We make only one exception to our local-guide rule. When a trip visits several different countries, the guide will be from the first country on the itinerary. We call this *continuous leadership*, and we implemented it after reading our post-trip questionnaires. Travelers grow attached to their guide and don't like to see a change just because they have crossed a border. We believe paying attention to what travelers tell us is crucial to creating excellent trips.

By hiring native guides, we broke the tourism model wide open. Our competitors thought we were crazy. Most of them still do; in fact, 95 percent of travel providers worldwide continue to use guides from the home country. Certainly, using American guides would be cheaper and easier. But the benefits of having local guides are tremendous: they know the language, they understand cultural cues, they have friends and family in the country, and they have personal stories to tell. They know where the cool places and the hot bargains are.

Did we make some mistakes along the way? Sure. We hired a group of Egyptian guides we had heard were amazing, but they were rigid in their guiding and couldn't deliver the type of unforgettable experience our travelers had come to expect from us. They dragged down our quality and our sales in the region, and it wasn't until we fired them and hired guides who were open to being flexible and to trying new things in their delivery of a trip that we began to rebound.

In the end, local guides deliver *unforgettable experiences*—and that contributes to the *unsurpassed value* of our products, which is one of our Extreme Competitive Advantages. We might be crazy, but we made the right decision back then, and it's a decision that has won us thousands of loyal American travelers.

Go Local to Deliver Excellence

Alan: Also in 1996 (our eleventh year of running Grand Circle), we discovered another way to improve on our goal of unsurpassed value and excellence. We expanded our global operation. It may sound strange, but we went global so we could become more local. We recognize this isn't the path for most organizations. Other companies go global to expand their markets, or to outsource their manufacturing, or to increase their visibility to foreign governments. We weren't looking to do any of those things. We went global so we could get in-depth, in-country knowledge to make our trips more authentic, more unique, more *life-changing*. We went global so we could get direct purchasing power to improve our pricing. And we did it so we could be right there when and if our travelers needed us. Going global would improve both the value we could offer our customers and the excellence of our trips.

We realized we would never gain control over the quality of our trips until we broke free of the self-serving ground operator system. We also realized we would never break free of this system from Boston because we didn't have the knowledge, the networks, or even the languages to do it. We needed to establish operations offices that were closer to destinations.

Bucking the system would be difficult—even risky in some countries. In many developing nations, the travel industry is tightly controlled and protected by the national government, and most countries require that foreign companies create a legally incorporated subsidiary before dealing directly with vendors. That's a lot of bureaucracy and paperwork. In many areas, deep-seated guilds control the assignment of tour guides to trips; we knew they would fight to distribute plum jobs to our big-tipping American tour groups. Ground operators have power, too; they can threaten to take business away from vendors who deal directly with us. In fact, this has happened to us many times.

With governments, guilds, and ground operators lined up against us all over the world, our office in Boston was overmatched. We were clear on our strategy. We needed to establish a local presence to increase value and excellence. But to do that, we needed to make another organizational change. We needed to open overseas offices and staff them with local people who knew the ropes.

To understand how enormous this change would be, it might help if we explained how we operated before 1996, and how most travel companies operate to this day. In those days, Grand Circle booked hotel rooms, rented motor coaches, and arranged meals through independent ground operators in the host region. The ground operator also coordinated with unions to get us tour guides. In other words, the ground operators acted as intermediaries to arrange all the elements of a trip. And they took a healthy cut for their services—up to 50 percent. In many parts of the world, there were layers of kickbacks and favoritism built into the system that increased our costs. Replacing underperforming vendors was slow and difficult. There were always excuses, excuses, excuses. In our view, the ground operators had way too much control over the trips. We needed to manage our trips ourselves.

We knew there was another way because we had already tried it. In an earlier chapter, we described our company's travel philosophy and recounted how Honey Streit-Reyes coined the term *unforgettable experiences* to describe the type of European tours she was organizing from her native country, Germany. Honey was our first direct buyer in Europe, which at the time was our primary sphere of operation, accounting for 75 percent of our traveler volume. Honey had already bypassed the ground operators in Germany, Austria, and Switzerland, which saved us a lot of money. She also improved our trips significantly by arranging for home-hosted meals, surprise excursions to out-of-the-way places, and unscripted encounters with local people. Honey's trips always scored high on our travelers' quality surveys.

In time, we came to understand that it wasn't just Honey. Local people always know their country best and always deliver the best trips. Today, it is obvious to us. Our local buyers negotiate face-to-face with our vendors, and they get better prices because intermediaries aren't taking a cut. Our local finance people resolve vendors' problems on the spot and in the native language; they also deal easily with local currencies. Local managers select and train guides better than we can from Boston; they also understand local laws and deal effectively with their governments. Local people make our operations run smoothly, and they help overcome inertia when problems arise. They are that important personal presence when a phone call, fax, or e-mail from Boston can be easily ignored.

Similar benefits appear on the product side. Local people make our trips less "touristy." They find family-owned hotels and restaurants that reflect their own culture; they know places where Americans don't ordinarily go; they find locations for our discovery programs; and they also design our travelers' visits to our foundation projects. Because they are on-site, they are the first to know when local conditions change. They know when train workers go on strike. They know when the quality of a hotel suffers under new management. And our local guides do a much better job explaining their culture and history to our travelers than even the most experienced American guides.

In 1996, the advantage of local control wasn't quite as clear to us as it is now, but as we looked harder at our quality survey data, we found that the more local people controlled the design and delivery of our trips, the higher the quality scores from every region. Locals know best.

Harriet: We had been considering going global for a while, but Alan made the decision during an off-site for the senior leadership team in 1996, in New Brunswick, Canada. We knew we had to change, or we would never achieve our vision to be the world leader in quality and value. Still, there was much discussion among the company's senior leaders about the pros and cons of this decision. Our vice president of operations worried that the change would drive up costs. The president of our Overseas Adventure Travel division worried that we would lose control of quality; OAT had just launched its first trips in Europe, and the president didn't want some relative stranger making decisions about her trips. Everyone was concerned about how we would monitor what the overseas offices would be doing. But Alan was sure we needed to cut out the intermediaries and be our own local knowledge. So we did it. We began building a network of overseas offices.

It wasn't a smooth build-out, by any means. The launch of the first office, in Munich in May 1997, proved to be expensive, contentious, and hard on morale in Boston. It was a high-tech office, with five desktop computers, four laptops, Internet access, and videoconferencing. The idea was to have Munich manage all our European products—which at that time numbered 35 trips with 1,000 departures a year—while Boston retained sales, marketing, finance, and customer service functions. Later, we brought accounting over to Munich, too . . . and then everything else. We had high hopes.

But there were glitches, and in retrospect, opening the Munich office was a mistake. The rent was high, the associates were sometimes too independent, and one even stole some money. We soon realized we needed many more than the 10 people we had staffed it with. And everyone in the Munich office resisted everything that came from the Boston office. In short, we had underestimated the difficulty of running an overseas operation, and we had moved too fast and without good controls.

But without taking this first step, Grand Circle would never have built our extremely strong and effective worldwide organization. We knew that locals know best, but it took us a while—and a few big-ticket mistakes—to learn how to direct a worldwide organization to take advantage of local knowledge and expertise. There was an upside: the Munich office showed us that local control of our program services, including hiring and training our own guides, could greatly improve our quality scores. It also helped us expand the practice of buying direct in Western Europe, securing our hold on our most important market. We had our sights on new markets, too, especially with our fast-growing OAT trips. Unfortunately, we soon discovered that the Germans were just as hesitant as the Americans to tackle the more remote parts of the world.

Munich taught us that if we were going to buy direct all over the world, we needed offices all over the world. In 1996, we already had about 30 people working for us in several countries overseas, but they worked on their own, mostly out of their homes, and their responsibilities were limited. Now we were thinking about an entirely different kind of overseas presence. We survived the problems with the Munich office, and we pressed on. Within two years, we were in Cape Town, Hong Kong, and Bangkok; then came Rome, Sydney, Paris, and Warsaw. By 1999, with footholds in Europe, Africa, and Asia, we were committed to building a geographically dispersed, multinational, and multicultural workforce of several hundred people.

Today, we have 34 offices in 31 countries on 6 continents, and we do business in 60 countries. We have more than 2,200 people working for us—in our Boston headquarters, in one of our regional offices, as ship crew members, or as guides. We are the most international company in the travel industry, a global organization that's committed to operating locally. Here's a list of our worldwide offices:

1. Arusha, Tanzania
2. Bangkok, Thailand
3. Basel, Switzerland
4. Beijing, China
5. Boston, United States
6. Bratislava, Slovakia
7. Buenos Aires, Argentina
8. Cairo, Egypt
9. Cape Town, South Africa
10. Chiang Mai, Thailand
11. Cusco, Peru
12. Delhi, India
13. Dubrovnik, Croatia
14. Guatemala City, Guatemala
15. Hanoi, Vietnam
16. Ho Chi Minh City, Vietnam
17. Hong Kong, China
18. Istanbul, Turkey
19. Killarney, Ireland
20. London, England
21. Luxor, Egypt
22. Lyon, France
23. Marrakesh, Morocco
24. Moscow, Russia
25. Panama City, Panama
26. Phnom Penh, Cambodia
27. Quito, Ecuador
28. Rome, Italy
29. San José, Costa Rica
30. Santiago, Chile
31. Siem Reap, Cambodia
32. St. Petersburg, Russia
33. Sydney, Australia
34. Tel Aviv, Israel

We made a lot of mistakes along the way. We trusted our overseas colleagues too much and gave them too much financial freedom. Two charming brothers from Costa Rica ripped us off for years before we discovered it, and a trusted and beloved leader in Croatia took kickbacks from vendors for giving them our business. We learned from these experiences (and many more like them) to rein in the financial controls and to monitor overseas activities much more carefully. But we also learned this from going local: it works.

6

Measuring for Excellence (Again and Again)

Ask your customers what they want— then listen and deliver.

*H*arriet: We believe the best way to grow your business is to please your customers. Not just serve them, *please* them. And we think the best way to make sure your customers are happy is to continually ask them what they think of your products—then *pay attention* and *use the information they give you!* So many companies don't know anything about what their customers really think of their products or services. Even companies that do customer surveys squander the information they get by not paying enough attention to it. Or they never do anything to change their products or services based on the information they get.

At Grand Circle, we want to know what all our customers think of our trips, and so we send every customer an extremely elaborate post-trip questionnaire. Grand Circle used customer surveys way before Alan and I acquired the company, but those surveys were nowhere near as extensive as the ones we use now. Plus, they were more qualitative than quantitative and so were hard to measure. Back then, in the early 1980s, the survey asked only 25 questions or so. Now our surveys are 12 pages long and ask 90 questions or more. (Appendix C shows a sample page.)

Here's how our surveys work. Customers returning home from their trip with us find a survey in their mailbox right when they arrive home. More than 70 percent of our customers respond. If you know anything about direct marketing, you know that's a *huge* number; some travel companies are happy to get a 3–20 percent return, and other industries that do other types of direct-response marketing or surveys are usually satisfied with only a 0.1 percent response rate. But it's our experience that people arriving home after an overseas experience are eager to tell us about their trip—the good, the bad, and the ugly. They not only answer the check-the-box questions; they also write copious longhand comments.

Our questionnaires cover everything from the customer's first phone call with our company to his or her arrival home. The questionnaires are customized for each trip and coded by departure date. That enables us to track the quality of the trip in different seasons and under different trip leaders and drivers. The questions probably look excessive to people accustomed to customer response postcards. But this information is our lifeblood: it literally determines how we manage future departures. Leadership in the travel business—and many other businesses—requires a vision, but it also requires *information*.

Most companies ask for feedback from their customers, but few of them act on it as obsessively as we do. When we bought the company, there was no technology to support the surveys. All we could do was read them and take notes. Now, we scan every questionnaire electronically. We tabulate all the quantifiable data. We create a database that is accessible to our associates all around the world. We read all the handwritten comments. And we send the results both to the regional office to which the trip is assigned and to the appropriate departments here in Boston.

Alan: Here's how we handle the volume of information coming in. We scan the surveys electronically; one person on our quality assurance team, Michelle Devine, does all the scanning. And we tabulate the results by description: *excellent, good, fair,* and *poor.* Hot issues go straight to the quality assurance team for resolution—and the team works closely with the regional offices, Worldwide Business operations, and other departments (sales, marketing, Web, etc.) depending on what the issue is—to mitigate or resolve the issue satisfactorily. Praise letters are shared with the regions, posted in the elevators in our Boston office, and highlighted in *Bridges,* our weekly e-newsletter.

We also create reports for each trip (for instance, the Great Rivers of Europe trip or the Heart of India trip), and we track the excellence ratings of each trip month by month. If a score goes down over the course of a couple of months, we go into the details to see what the problem is. Everyone in the company can see the results of every trip and every departure, and everyone has access to the individual surveys. All this information is on the Grand Circle intranet. Primarily, though, people in the relevant office or department look most closely at the surveys. They're watching for trends, hot issues, and the performance of our guides, while the quality assurance team is looking for hot issues (overall) and for specific traveler complaints to see if those complaints are part of a negative trend.

We tabulate the results, but until recently, we weren't able to contact every passenger who reported an issue with a trip. In the past, we had to tell our travelers that because of the volume of responses we received, we could not answer individual comments on the surveys, and so we asked that anyone who had a pressing issue call, e-mail, or write to our quality assurance team. In the past year, though, we've hired someone to read the surveys for hot individual issues: Barbara Harrington is now part of our nine-person quality assurance team, and she brings to our attention the hot issues so we can discuss them and then contact the traveler to see what we can do about that particular issue. The head of our quality assurance team, Diane Simpson-Pye (who has been with us for 19 years), tells us that most of the travelers we contact personally are incredibly impressed that we are calling them, and they are very surprised by our proactive approach when we've told them to take another step to reach us with specific issues.

In another effort to provide high-touch service to travelers and to foster long-term relationships, we launched a Customer Loyalty Team—now consisting of three people but soon to expand. This team looks at travelers who gave us only "fair" or "poor" ratings on their surveys, and then the team contacts those travelers to find out what happened and to see what we can do to encourage them to give us another try. These calls differ from the quality assurance team's calls to individuals with specific issues because they address a traveler's overall low rating, rather than something specific.

We also launched "Harriet's Corner" on our Web site about three years ago. The section on our Web site shares travelers' stories, news from us, and personal updates from Harriet. Harriet was really involved at the outset and responded to all the e-mails that came in. The volume grew so much, though, that we hired a writer to help her keep up. The volume continued to grow, and so we now have a dedicated associate assigned to monitor "Harriet's Corner": she sends issues to the quality assurance team, responds to general queries, and forwards to Harriet e-mails that she needs to know about. The system works pretty well, and, eventually, we are going to make "Harriet's Corner" interactive so that our travelers (who are mostly women on this part of our Web site) can communicate directly with each other.

In addition, although we're way behind in terms of social media, we're catching up. We now have Facebook pages for each brand (GCT and OAT) as well as for Grand Circle Foundation, and we have a dedicated monitor and writer-respondent for each. We feel that Facebook, for example, will give travelers a new way to interact with each other and give us another way to build our traveler community. Our new social media manager just started in June 2011, and we hired her boss only four months earlier; so we have a long way to go, but we're excited about the possibilities.

Anyway, back to pleasing our customers. When a consistent problem arises or when a pattern emerges from what we read on our customer surveys, our regional offices immediately work to develop an action plan to address it. For example, several years ago, we were offering a trip to Israel, a popular destination for many travelers—Christian, Jewish, and Muslim alike. Unfortunately, the trip tanked in our travelers' quality surveys, garnering only a 46 percent "excellent" rating. Our travelers let us have it. The guides were terrible, they told us; the hotels only so-so. What's more, the trip had no soul.

Our travelers are all about relationships—relationships with their families, with their friends, with their trip leaders, with each other, and with us. Ordinarily, we design our trips with relationships in mind. But this time, we had failed. Our travelers wanted to have more personal and more familial experiences in Israel. They wanted to get behind the often-tragic headlines and make a more human connection with the people of the country.

We listened carefully to their suggestions. Then we made several important changes to the trip. We terminated our ground operator and hired local staff. We hired and trained our own guides. And we added many more cross-cultural encounters. We included a dinner within the Jewish Orthodox community. We arranged a discussion with the leader of a Bedouin women's society. We set up a meeting with Jewish farmers. We organized a talk with a Palestinian woman. And we included a home-hosted meal with a Druze family.

What happened? Well, our "excellent" scores rose from 46 percent to 82 percent in a few short months. Once again, the fact that we often make mistakes, but that our travelers will always set us straight, rang true.

Once a team of outside observers characterized us as being "maniacs on excellence." They were right. We believe if you want to build a great company, you have to ask your customers how they feel about your products. You have to listen to what they have to say. And then you have to act on what they say. That seems obvious, but many businesses don't ask their customers for feedback. Or they don't act on the information when they do.

In our case, we generally believe that all our trips are excellent, but as with our Israel trip, our customers have told us otherwise again and again. Fortunately, in addition to pointing out our errors, they also tell us how to improve. We change trips all the time in response to what travelers have told us. For example, on our Panama trip, we recently took our travelers' advice to move the transit of the Panama Canal to the end of the trip, to serve as the climax of the adventure.

We also respond quickly when our customers tell us we've made a big mistake after we've changed something they've liked. For example, when the dollar weakened dramatically around 2005, we took some included features out of our trips and made them optional excursions. This strategy allowed us to keep our base prices low, because optional excursions are priced separately. But it hurt our quality scores by undermining two of our Product Pillars: value and discovery. When we saw the results, we put these excursions back into the base trips. Like magic, the scores went back up.

This was a big mistake because we tampered with one of our Extreme Competitive Advantages—*delivering unsurpassed value to our customers,* which is the foundation of our product strategy and the key to our dominance in the industry. Lesson learned. We need to stay with our Extreme Competitive Advantages because they're what has made us successful and what will set us up for future success.

Over the last five years, we are proud to report that we have raised our overall excellence rating 5 points, to 81 percent. We've done it simply by listening and responding to our customers. Good leaders listen. They admit when they make mistakes. And they fix them. Our experience is that customers know what they want. When we deliver, our customers become unassailably loyal to the company, traveling with us again and again.

Listen Closely to Your Customers— They're Your Best Consultants

Harriet: Some companies ignore what their customers tell them if they think they know better or know more than their customers do. Of course, you know your business. But when it comes to an honest evaluation of your products, your customer knows best. We learned this the hard way, a few times.

For example, years ago, our Russian river cruise tour came up short on quality scores. A big problem was the food. The travelers said it was awful. It was "boring" and "repetitive," and it "had way too much starch." You might laugh at these comments, but people who don't enjoy their meals on their trip typically don't enjoy the trip too much either.

Another problem was that two of our senior leaders didn't really think the trip was all that great. They were concerned about the quality of the ship, which wasn't as modern or as nice as our European river ships. Although our associates in Moscow and St. Petersburg said they were making progress, we canceled the product. Why? Because we believe in cutting products that don't meet our standards. And our big concern was that if the trip didn't meet expectations, we would lose loyal travelers. We couldn't risk that.

Cutting the trip turned out to be a big and costly mistake, however. The quality scores for the final trip departures actually *exceeded* our goals. But it was too late. We had already canceled everything: our airline seating holds, hotel rooms, coaches, restaurant reservations, entrance tickets to museums and other sites, everything. Our travelers were right all along—the food was boring, but then it got better. But we acted too fast in canceling the program.

We had also made another mistake. We failed to listen to our associates in Russia. They had the most current information. They had their eyes and ears on the ground. We lived to regret it, because it took us five years to get back into Russia. All that time, we kept kicking ourselves, saying: *Listen to our customers and our local associates on the ground.*

Word to the wise: you need to do the same in your business!

Talk to Your Customers Personally

Alan: Maybe you can't get your customers to fill out customer response cards or surveys. Maybe you can't get your customers to talk to market researchers doing phone surveys. Don't give up. You still *can* go out and talk to your customers face-to-face and ask them what they like and don't like about your product or service.

We did that right from the beginning in the fall of 1985. We went into the field. We checked out the accommodations. We took the tours. We ate what our customers were eating. And then we held cocktail parties offering free drinks so we could meet with people and talk to them.

Face-to-face human interaction with our customers has always been our model. We believe you don't know what people really think until you talk to them. First, I took a few trips. Then our first president, Bruce Epstein, took a 3-week whirlwind trip in November 1985. Then all the execs started going in groups of four to five people each, for 10 days or 2 weeks. These weren't leisurely trips; we were banging through different cities and countries to get a taste of the product line and meet as many travelers as we could.

In fact, that was a big part of Mark Frevert's job for about 15 years. From 1985 to about 2000, Mark met with tens of thousands of travelers. One memorable event took place in Torremolinos, on the Costa del Sol in southern Spain. We had asked the local staff to set up a cocktail party and invite all our travelers. Unfortunately, the woman who ran the program was very angry about the recent changeover in management. So she put up a sign saying, "Executives from the Boston office will hear your complaints at 6:00. Free cocktails."

When 450 people showed up for the free drinks, about 150 of them had complaints. They complained about everything. "Breakfast doesn't start early enough." "My plane is routing through Frankfurt, and I'd rather transfer in London." "My hotel room is too small." Mark was in the hot seat, but he took off his suit jacket, loosened his tie, pulled out a notepad, and started jotting down the travelers' concerns. He was there for hours, but he listened to everyone who wanted to talk to him.

Getting the specifics from customers is invaluable. If listening to 150 people complain is what it takes, you should thank each and every person for taking the time to tell you what you can do better. Remember, the alternative is *they might stop being your customers!*

We also hold "traveler appreciation" events around the country. We usually have these at a local hotel. We include breakfast or lunch—people are much more likely to come if there's food. These events are a way to recognize our best travelers. They are also a great way to meet potential new customers, because many travelers bring their friends. And they are a way to update our travelers about new trips, added features, and top travel destinations. Most important, though, is that these traveler appreciation days give us a wonderful opportunity to hear from travelers face-to-face. They tell us what's working—and what's not.

For example, between January and June of 2011, we held 42 events in 12 states—with more than 19,000 attendees. That's a lot of people offering feedback—and we appreciate every bit of it.

Build a Big Database

Harriet: Today, our customer database contains a tremendous amount of information about our customers' travel preferences. We know where they want to go. We know where they don't want to go. We know how much they spend on average on each trip. We know whether they take a pre-trip or post-trip extension trip. And we know a host of other facts and figures.

Alan, Mark Frevert, and others in the company had worked in direct marketing, and so we've always known how to apply direct marketing to the travel business. In our industry, a lot of rules of direct marketing don't really apply, because travel is such a high-ticket item. It's not like we're L.L.Bean or other retailers that mail out a catalog and get a high response rate, but then the average sale is only $50 or so. Travel is a different animal from that model: our average ticket for a couple traveling is about $10,000.

Also, we use a two-step marketing program, whereas other industries that do direct-response marketing use large lists of potential customer names and target different segments of those lists. In contrast, our two-step model involves placing blow-in cards, for instance, in a travel magazine. These cards wouldn't try to sell a trip. Instead, we begin with a free catalog offer. And we ask only a few questions on the card. We ask people their age. We ask when they last traveled abroad; e.g., "was it 3 years ago, less than 3 years ago, more than 3 years ago?" And we ask them what areas of the world they might be interested in traveling to.

That is step 1 of our model. Step 2 is to follow up and send a catalog of our trips to people who send in the response cards. Here, too, the travel industry model of direct marketing is different from that of other industries. Typically in direct marketing, if you haven't sold someone a product within the first six months, you're probably going to stop marketing to that person. But because travel is so expensive, we found that people who continued to get catalogs over time would eventually travel with us. At Grand

Circle, we never delete people from our database (unless they ask us to). We may contact them less frequently. But we don't eliminate them. We believe that sooner or later, they're going to travel with us. And once they do, we're confident they'll travel with us again—and they'll recommend us to their friends and family; 24 percent of our business comes from referrals. Our database now consists of 5 million U.S. households (50 and older) who have expressed interest in foreign travel.

Alan: In addition to our database, we've conducted research studies to find out what our customers want from their travel experiences. Mark Frevert handles these studies, which we do every few years or so.

The first research study Mark did was in April 1986, only six months after we acquired Grand Circle. Mark hired an outside research company to do an extensive survey of about 320 questions. We mailed it under the research company's name, so that the people being surveyed wouldn't know that our company was behind it. (Some of our customers probably guessed we were the sponsor, especially if those customers had traveled only with us.)

We pulled a variety of names off our database to survey. We sent surveys to both men and women. We sent them to people who had taken a lot of trips, to people who had taken only one trip, and to people who hadn't taken any trips yet. Our first survey also included a large demographics section, about three to four pages, which asked all the classic demographic questions. We asked respondents' age, marital status, and number of children. We asked their household income and total net worth. We asked about their education, military service, and careers. We asked whether or not they owned a second home. And so forth.

The research company we hired mailed that survey to about 12,000 people. We got back a 74 percent response! That totally blew away the researchers: they had never seen a response anywhere near as high as that. But it didn't surprise us. We know that people who travel are hugely passionate about travel. Asking them about their experiences isn't anything like calling people and asking them how they like their office copy machine or what brand of toothpaste they use. When you ask that type of mundane question, most people won't be as interested in talking to you.

This 74 percent response rate was especially impressive because 90 percent of U.S. foreign travel is to Canada, Mexico, and the Caribbean. In other words, very few people travel to the far-flung places that Grand Circle travels to. In fact, only 37 percent of Americans even have a passport.

That survey, in 1985, cost about $28,000. But we didn't balk at the price, even though we were still losing $2 million a year (or $5,000 a day, as we thought of it). We didn't have money to waste. But this was money well spent, because the information we learned from that survey was terrifically extensive.

Again, you need to find out what your customers want, and we truly believe that the best way to do that is to *communicate directly with them*. It's worked for us, and we believe it can work for every company and in any industry.

We still do the occasional professional research study. Every five years or so, we do a benchmark study. We don't need to do them more often than that, because in the travel industry, the trends are glacial. They're not fast moving, as they are in other industries, such as technology, for example.

For companies whose customers aren't passionate about their products or services, as travelers are, you can still survey your customers. But as mentioned, it's harder to get them to respond. Still, one of the easiest ways to invite response is to offer an incentive. We've done this as well. We've held a sweepstakes where we offer the winner a free trip for two to the South Pacific, which is an $8,000 value. One of the more effective approaches we've seen is to put $1 in the outgoing envelope with the survey. You won't find a lot of companies doing this, because a lot of those dollars are going to end up in the garbage because people won't even open the envelope. But for those who do, there's a huge guilt factor to complete the survey. There's always a way to get people to talk to you about your product. You simply need to find out what's important to them to get them to open up.

Linking Customer Satisfaction to Compensation

Harriet: Every organization uses financial and operational measurements to track how well they're doing. However, few organizations make customer satisfaction their *primary* criterion for success. In most organizations, employees get evaluated and compensated on whether sales go up or down. But they aren't evaluated and compensated on the basis of how well they have delivered excellence from their *customers'* perspective. In fact, most employees in most organizations would balk at such a metric because it's so subjective. Most employees would much rather be evaluated and compensated on something tangible. Like how many products they sold. Or how much revenue increased. Or how many marketing campaigns they managed. And so forth.

Measuring employee performance on the basis of customer satisfaction, however, is a much more compelling metric than the rear-view mirror metrics of "What happened to our sales?" That's a result of what your company did two years ago. It's not a result of what you're doing *today*. The fact that we focus not only on measurement but on measurement of customer experience is unusual. Even more important and unusual, though, is the fact that those measurements are linked *directly* to compensation and bonuses.

On every trip, every activity has an excellence target. We don't rate only the obvious things, such as the quality of the food or the hotels. We also rate the ability of the tour (and tour guides) to provide our customers with an experience of learning and discovery. If the tour guides only show our customers the easy stuff, then they haven't done their job.

That's why our "free passes" are so important. We need guides to be constantly experimenting with what will make every trip a more interesting learning and discovery experience. But because the guides are compensated based on how well they do, there's a tendency among guides to keep doing what they know will be successful, rather than trying something new. So the free pass provides a way where, once or twice a year, our guides can say, "I want to try a new experiment with this trip." And that experiment doesn't count on their performance evaluation score. In fact, some area managers actually *require* their guides to try a free-pass activity every year. Because if they don't, they're not experimenting. And if they don't experiment with something new, the trips will become stale and boring, and our travelers' satisfaction will diminish. We want to prevent that.

Also, we don't hold all guides to the same standards for all trips, because some destinations can only offer so much, and it would be unfair to penalize the guides for not delivering something that simply can't be delivered, no matter how fabulous the guides might be. For example, our trips to Vietnam get incredibly high excellence scores. There are so many interesting things there to show travelers that they've never seen before. On the other hand, it's much harder to get excellent scores on some of the South American and African countries. Why? Because the hotels aren't as good overall as they are in, say, Europe. So those trips have different targets.

We also realize that "excellence" is relative to customer expectations. For example, we have brand-new ships that offer extraordinary first-class programs, and they cost more than some of our other cruises. About 90 percent of our travelers rate those trips as excellent. Then we have other, less expensive trips on older ships that are not as fabulous as the new ships. The customers who go on those trips know they're going on older ships. So we get only 80 percent excellent ratings on those trips. We can't expect to get 90 to 95 percent excellence ratings on these older ships, and so the target for the trips on these older ships is only 85 percent. If one of our new ships got only 85 percent, we would wonder what was going wrong. And you can bet that we would be moving quickly to find out and correct it!

Measure Your Managers' Performance

Harriet: Finally, one other measurement that deserves mention is the performance of managers. Grand Circle asks every one of our associates to evaluate their managers and their leadership teams. These are not anonymous evaluations, and so some businesspeople might think that could never work. Many people think our associates would be afraid of expressing their viewpoints candidly. But it does work. It works because at Grand Circle two of our six core values are open and courageous communication and risk taking. Our associates know we expect them to be open and to speak up even if it's a perceived risk. We're not afraid to hear from anyone how we can do better. We want to—we need to. Our goal is to improve. All the time. So we appreciate any and all feedback we get from anyone.

7

Fostering Customer Loyalty and Building Community

Take a "high-touch" approach to customer service.

*A*lan: In today's economy, many businesses think the road to success is through cost cutting. Companies look at automatic teller machines, pay-at-the-pump gas islands, and self-checkout grocery stores; then they try to figure out the next service they can get customers to perform for themselves. Wages are lower in Asia, and so many companies outsource their customers' telephone calls to India. Or they send their manufacturing plants to China. With the exception of Apple, it's nearly impossible to physically locate the company that built your personal computer. And just try to find help in a department store. You can wander around for five minutes looking for someone to unlock the dressing room and ring up your purchases.

American business is going "low touch." It's a business strategy driven by bookkeepers, and it leads to a pretty cold and calculating appraisal of the customers' value to the company. In fact, many companies feel no need to build a relationship with their customers at all. For many businesses, a one-time customer is good enough. They believe *price* is the biggest factor influencing consumers' choices. So they cut, cut, cut their costs and pricing—all the way to the bone. They think investment in customers and customer service is a waste of money because when price is the driver, customers will go down the street for their next purchase anyway.

That business model doesn't make sense to us. We're not selling widgets, or hamburgers, or sacks of corn. So we don't think of our trips as "commodities." We think of our trips as *experiences*—exciting and unique adventures that have the power to make people's dreams come true. Our travelers become emotionally involved with our trips. They enjoy sharing dim sum with a family in rural China. They enjoy playing tug-of-war with schoolchildren in Tanzania. Those activities have the power to astonish our travelers, to teach them, to move them, and to change their perceptions of the world. Both the trips and the travelers require special handling.

At its best, travel is a people business. It's a business that benefits from a close, two-way, high-touch relationship between a company and its customers. This is the kind of relationship we have built with our travelers, and it has become one of the foundations of our business. It not only guarantees repeat business for the company; it is also a source of personal satisfaction for Harriet and me. And apparently, for our travelers. Consider the words of Charlotte Gates, one of our many repeat travelers, who wrote this to us: "Never mind that I have spent my children's inheritance with you. I have walked the Great Wall, viewed the Taj Mahal, and ridden a camel to the Pyramids. You are *my* company."

Of course, travel is not the only people business that benefits from high touch. There are countless others. Any business that sells expensive products or services benefits from a high-touch approach. After all, you wouldn't buy a Bentley over the Internet. Instead, you'd go to a dealer you've bought from before who would find the perfect car for *you*. Moreover, Starbucks proved you can decommoditize anything, even the lowly $1 cup of coffee. If you make your brew exotic and the ambience welcoming, customers will walk right by the corner diner or deli and gladly wait in line for 15 minutes to pay $3 for a medium latte, just how they want it—especially if the barista remembers what they order when they walk in.

We spend a lot of time and a lot of money listening to our customers in order to measure and improve our performance. Listening creates loyalty. Here's what's worked for us—and what hasn't. I hope it can help you develop greater loyalty from *your* customers.

Know Your Core Customers

Harriet: Earlier, we described the first professional research survey we conducted after acquiring Grand Circle back in 1985. That survey went out to 12,000 people from our database. It asked all the conventional demographic information. And it provided us with a basic snapshot of our core customers:

A SNAPSHOT OF OUR TRAVELERS

> Almost all are over age 50.
> 65 percent are retired.
> Their average age is 70.
> They come from all 50 states.
> 35 percent have a background in education.
> Two-thirds are women.
> 30 percent travel solo.
> 55 percent have traveled with us before.

This is just the basic information. We have lots of other facts and stats that give us an extremely detailed understanding of who our customers are and what they want when they travel. This information helps us give customers what they want. For example, two-thirds of our customers are women, and they told us they wanted to find other like-minded women to travel with, and so we established our Travel Companions program. This is a free service on our Web site that helps travelers connect with fellow travelers as possible travel companions for an upcoming trip. We launched a matching roommate program for same sex travelers. And we promised that if we could not find someone a roommate match, then the person would get a room without having to pay the dreaded single supplement fee.

Repeat Customers Keep Marketing Costs Low

Harriet: Most of our customers take their first trip with Grand Circle because a friend has referred them and they like our low prices. But they *return* because of the value and the quality of the experience. They discover that we go to interesting destinations and that we employ great guides. They discover that we use centrally located hotels, that we visit local villages and schools, and that we include more meals and features than do our competitors. We like to say we deliver a four-star trip at a three-star price. We miss occasionally, but when we do, our travelers let us know. And we use that feedback to improve.

Our business model depends on these returning travelers. *Acquiring new customers is very expensive,* especially in the travel business, because trips are such big-ticket purchases. The same is true for other industries selling expensive products or services. It costs us about 20 times more to book a new customer than it does to book someone who has traveled with us before. We've found that the least expensive way to fill our trips is with people who have already traveled with us. Repeat travelers know the ropes. They share a certain sense of camaraderie. And they help make our trips more enjoyable for the first-timers. They also talk about their previous trips among themselves—in fact, bragging rights are a big part of mealtime conversations. And many decide where they will travel next based on these conversations. Happy travelers often return home and refer their friends to us, too. This is an added bonus. In fact, *80 percent of our profit from first-time travelers comes from these referrals.*

Here's what our model looks like: repeat customers keep our marketing costs down . . . which helps us keep our prices well below the competition . . . which in turn keeps our customers coming back. It's a circle. And it drives both the company's profit and our customers' satisfaction.

Repeat business isn't the only thing that keeps our costs down, of course. In addition, our worldwide organization, our "buy-direct" strategy, and our targeted marketing all play a part. Together, they create our *unsurpassed value* (one of our Extreme Competitive Advantages). Our strategy is to pass our savings on to our customers in the form of lower prices. Then we deliver surprising value once they depart on one of our trips.

Why do we do this? We do it because of the circular nature of the business model: if you break any piece, the model falls apart. We could jack up our prices and make a ton of money—for a short period of time. But then we would lose the volume from returning customers. And our marketing costs would escalate. Lower volume would also upset our vendors, who would start raising their prices to us. Our costs would rise. And pretty soon, we'd be like every other travel company.

Our unsurpassed value model works exceptionally well. In good times, we grow in customer volume and profits. In tough times, our loyal customers keep the wolf from the door. More than once, they have kept traveling with us when other travel companies have seen their business stop cold. For example, we were the only U.S. tour operator in Egypt after 9/11. Everything depends on this win-win relationship with our customers. That's why we put a lot of time, energy, and thought into it.

How to Get Repeat Business

Alan: The relationship we have with our customers amazes us. They constantly tell us that Grand Circle is *their* travel company. And the facts support this:

- 55 percent of our travelers are repeat customers.

- More than 33,000 households have taken 7 or more trips with us.

- One couple, the Youngs of Jacksonville, Florida, has gone on *61 trips* with us over the past 21 years.

- 99 households have 30-plus trips with us stamped on their passports.

So how do we do it? We believe there are several key building blocks to developing customer loyalty.

Respect Your Customers and Ask Them What They Want

Alan: The trips are the big draw, of course, and we work hard to deliver great value and unforgettable experiences. Ralph Moody, the Hall of Fame race-car driver and race-car builder, once said, "If you make something that's good, you're going to sell the hell out of it." He's right.

What's different about how we do it is that we give our travelers a big say in how we design and deliver our trips. As we described earlier, we listen to every particle of advice they give us. We read their letters. We read their quality surveys. We listen to them at cocktail parties. We listen to them on the telephone. We read what they e-mail us. And we listen to them at special gatherings. Then, we act on what they tell us.

We believe our travelers know best, and we try like crazy to give them what they want. Maybe you're a veterinarian caring for your customers' beloved pets. Maybe you're a tailor who needs to make sure your customers look good. Whatever your business is, you need to listen to what your customers want. After all, no matter how good you are, there's always another vet, or tailor, or travel company that your customers can turn to. You may think you're unique. But if you don't listen to and care for your customers, *they won't listen to you or care about you either,* and they'll go elsewhere. *Don't let that happen.*

Offer Incentives to Encourage Repeat Business

Alan: Another important building block for us is benefits. We run a number of discount and loyalty programs to recognize and encourage repeat business. We offer a 5 percent discount to frequent travelers and a $100 credit for every new traveler our customers refer to us. (The new traveler gets $50 off that first trip as well.) We also formally recognize frequent travelers before and during a trip, with a phone call, cocktail party, and sometimes even a special gift presented publically. Whenever possible, we give them the best hotel rooms and ship cabins.

Combine unsurpassed value with good discounts and incentives, and our competition has a hard time luring our travelers away. This is such an easy benefit to offer. Yes, it costs money. But that money is a pittance in the grand scheme of your business. Why do you think companies like Direct TV advertise that they'll give you $100 and your friends $100 each if they sign up based on your recommendation? Why do you think all those "welcome to the neighborhood" packets include coupons from local dentists, doctors, home furnishing stores, and home repair businesses? They know that just 5 or 10 percent off your first visit can make you a customer for life. A little incentive goes a long way to engendering goodwill—and good word of mouth.

Be as Flexible as Possible

Alan: Another building block is travel protection. We know our customers lead busy lives that sometimes entail last-minute changes in plans. We also know people get sick or need to tend to family emergencies. These uncertainties often keep older people from traveling. They hesitate to make reservations, fearing they will lose their deposits or even the entire price of their trip. But not at Grand Circle. We understand that life sometimes throws curveballs.

In 2001, for example, after the terrorist attacks of 9/11, we became the first company in the industry to offer travel protection that gives a full refund in money or travel vouchers up to the moment of departure. Admittedly, it's a challenge to cancel flights, hotel rooms, meals, and excursion costs when someone cancels. But offering this peace-of-mind solution to customers has kept bookings high, even during tough times. *And* it's a terrific option for customers. It's a win-win for them and for us.

Harriet: Unsurpassed value, significant discounts, preferential treatment, and comprehensive travel protection make it easy and risk-free for our customers to book another trip. That's good customer renewal practice. It's also a good deal for both the company and the customer. But it's not really a relationship, is it? To build an honest relationship with customers requires another crucial element: interesting, consistent, and respectful two-way communication through which you *keep your customers informed* and *ask them what they want.*

Know What Your Customers Are Dealing With

Harriet: By constantly speaking with and listening to our customers, we've received early warnings about things that have gone awry—not only with our trips, but also in their lives. For example, in 2008, we began to feel the pressure of the oncoming worldwide recession. So did our customers. Many of them told us on their post-trip surveys that they still wanted to travel, but they needed some time to wait out the economic downturn. Others said they were holding their own financially, but they were reluctant to commit to a trip in case they might need their money later. People were simply more cautious financially.

What is remarkable is that our customers *told* us all these things. They didn't just turn their backs on us. Instead, they reached out to express their regret and their hope that they could travel again sometime soon. We understood their concerns. And we appreciated their confidences.

Of course, we also wanted them to keep traveling, and so we changed a number of our policies to make it easier for them to do so. We offered a temporary 30-day risk-free guarantee, whereby they could cancel their reservation. We offered free domestic air travel. We offered free single supplements on land trips and extensions. Every change either was a suggestion from one of our customers or was inspired by their feedback. They asked, we delivered, and they kept traveling.

Lowering prices or making other changes to ease your customers' financial burden is never easy for any business. After all, *you* need to survive and make a profit, too! But wouldn't you rather continue to serve your customers in *some* way rather than lose them altogether? We believe that good customers, loyal customers, will appreciate your efforts on their behalf. We believe they will stick with you when things turn around for them. In fact, they'll likely become even *better* customers in the long run.

Alan: At the beginning of this chapter, we mentioned that one of the key benefits of developing customer loyalty is that it keeps marketing costs low. Because 55 percent of our travelers are repeat customers, we really don't need to do much marketing to them, other than letting them know what new trips we're offering. However, we still need to acquire new customers—that other 45 percent of our customer base. So Chapter 8 takes a closer look at some of our marketing strategies to grow the business.

8

Creating Marketing Strategies to Leverage Extreme Competitive Advantages

Consistently communicate your Extreme Competitive Advantages to customers using both traditional and nontraditional channels to reach them.

*H*arriet: As we said earlier, one of our Extreme Competitive Advantages is how we keep our organization focused on the right strategies. We don't compete in areas where we don't think we can excel, and we don't waste money on tons of marketing approaches, focusing instead on exactly what works for us. This chapter covers some of the ways we focus our marketing strategies and our marketing techniques.

In a Global Market, You Need to Have a Differentiated Service

Alan: We promise to provide our travelers with a deep cultural experience, rather than a flyby of major tourist sites. We have a saying at Grand Circle: *go where Americans don't go*. It's part of our discovery formula, to visit places that other American companies don't usually take their travelers. For example, we visit the City of the Dead in Cairo, view funeral pyres in India, and explore Amsterdam's famed red-light district at night. We visit an undeveloped section of China's Great Wall where no other tourists are present. We include unusual modes of travel on our trips, too, like riding on camels, going river rafting, sailing on sampans and feluccas, and taking an overnight train ride in China—all typical for locals, but not part of most American travelers' experience. We go *off the beaten path* whenever we can. We get around as the locals do.

We believe this approach to travel is one of the things that sets us apart from other travel companies. We create a customer experience that's different. Although your industry is probably different from ours, we believe creating a unique or different experience for your customers can give you an advantage over your competition.

What are *you* doing to set your business apart from your competition? Whatever it is, it should be something that makes your customers return . . . again and again.

Help Your People Take Risks to Push the Boundaries of Convention

Harriet: We train our guides to raise sensitive issues that other travel companies consider "nondiscussable"—and get our travelers talking about them. In Vietnam and Croatia, for example, they discuss wars involving American and UN troops. In India, they talk about the caste system. In China, they talk about Tibet and censorship, and in Egypt, they discuss the role of women in a Muslim society. Americans want the straight facts, and that's how we present every country we visit—the good, the bad, and the ugly.

Alan: We also train our guides to literally "stop the bus." If our guides see something interesting happening on a walking tour or bus excursion, they will ask travelers if they would like to stop to watch or participate. Guides have stopped the bus for parades, political demonstrations, roadside craft sellers, cattle auctions, wedding processions, and street performers; they have even stopped the bus to help out with a rice harvest. This kind of ad hoc discovery event has the double benefit of showing our travelers something unexpected and completely unscripted and giving the guide something new to talk about. The trick is to seize the moment—carpe diem!

Although it's easier to simply follow a plan or program, especially if it has worked before and satisfied your customers, we advise against getting stuck in your usual itinerary or schedule. Most people don't want to feel as if their experience is the same as every other customer's. People want to feel special even if they are buying the same style of boots that everyone is wearing or eating at the same hot restaurant that everyone in town is trying to get into. They still want to feel they're getting something *different* in some way. So try to make each customer's experience as unique and special as you can. How about a personalized e-mail from your CEO after a customer purchases your product for the second time? How about a small discount, or free movie passes, or something that says, "You are special to us"?

Offer Genuine Experiences with Human Connection

Harriet: One of our programs is called A Day in the Life, and it has been a popular feature of many of our trips since 1989. The program is different with each trip and each group of travelers, but it usually involves spending a full or half day in a small, out-of-the-way village and experiencing local life firsthand.

For example, in Thailand, travelers walk through a village, meet the principal of the elementary school supported by Grand Circle Foundation, learn a simple dance from the children, tour the village's new agricultural cooperative (another Grand Circle

Foundation partnership), and try their hands at making traditional bamboo baskets and grass brooms. In Mexico, travelers visit a Mayan home and learn how to fry tortillas.

In each case, the Day in the Life program engages our travelers in relaxed, hands-on cultural exchanges with the community at large. It's an experience they can't get with anyone else, and it is part of our identity, our brand.

These village programs are tough to design and deliver, and we have spent thousands of hours on them. Making them an included feature imparts value to the trip (remember, delivering *unsurpassed value* to our customers is one of our Extreme Competitive Advantages); they also help vary the pacing of our trips. And although the day is structured, the program leaves plenty of opportunities for travelers to choose whom they would like to talk to and what they will do. Most important, these village-based discovery programs are enthusiastically received by our travelers, who invariably rate them very high on their post-trip evaluations—and they are a boost to the local economy, which supports another one of our Extreme Competitive Advantages, of *giving back.*

Offer a World-Class Experience to Customers Who Value What You Do

Harriet: All this emphasis on discovery and unplanned detours is second nature to Alan and me. As a family, we have always been adventuresome travelers. Small-group, wilderness-based adventures thrill us, and our travel journals are full of many fun adventures. Alan and I went trekking together in Nepal for his fortieth birthday. We took the kids hiking in the Costa Rican rain forest and rafting on Africa's Zambezi River from the base of Victoria Falls—a very adventurous trip that involved whirlpools, an overturned raft, alligators on the shore, and an angry me for Alan's subjecting his family to such a high element of risk!

So it is perhaps not surprising that we bought a company called Overseas Adventure Travel in 1993. There were strategic reasons for the acquisition, of course. We wanted to broaden our traveler base: baby boomers were seeking big adventures, and we needed to build a roster of trips outside of Europe and the Middle East, where American travel had been curtailed by the first Gulf War. But the reasons were equally personal: for us, "adventure travel" is pretty much synonymous with "unforgettable experience."

There were difficulties. The small, Cambridge, Massachusetts–based company was losing money, and its trips were too fast-paced and expensive for our market. Our strategy was to apply Grand Circle's four Product Pillars to the design and delivery of the OAT trips and then watch the quality reports like hawks until we were sure the trips suited our travelers' wishes and met our high goals.

The biggest challenge—and greatest opportunity—arose from OAT's signature feature: small-group travel. Our traditional Grand Circle land trips have an average group size of 38 travelers, a number that fits comfortably into a standard-size bus, but OAT land trips take only 10 to 16 travelers per trip. Small groups offer a more intimate travel experience, and they make it easier to get to out-of-the-way places; the downside of that is it's hard to make money on them because their high costs have to be spread over just a few travelers. So it was a challenge to make our value proposition work for small groups, but we figured out how to do it by making the trips longer (to bring down the per diem cost), dealing directly with local vendors in the destination countries, including airfare in the package price, and focusing our marketing on the unique and unforgettable experiences. We were offering what no one else was—niche market strategies all the way.

Small-group travel really does change one's travel experience. Small groups have access to restricted sites like antiquities that have been jeopardized by too many people tramping through them over the years. Small groups can stay in family-run inns and eat in local neighborhood restaurants that can't accommodate large tourist groups. Small groups create a special camaraderie among travelers, who often make friends with each other during the course of a trip and who may travel together later on. They connect more easily with children during school visits, engage in longer and deeper conversations at our Grand Circle Foundation sites, and get more individual attention from guides.

But as every adventure traveler knows, the "unforgettable" part of adventure travel is the adventure. Once you've slept in a tree hut under a starry Botswana sky, or sampled wild rat with villagers in Thailand, or pulled a new friend under a towering waterfall in Peru, you'll never be the same.

Our customers love these small-group tours. We know this not only because they tell us so in their post-trip evaluations but because OAT has been so successful. We believe that a great company does a few things well, and so when we make a new acquisition or start a really big initiative, we take care not to reinvent ourselves. We look to our leaders to keep us focused and disciplined, remembering our mission, vision, and values and applying our proven strategies to the new opportunity. We stayed true to our winning formula, rigorously applying Grand Circle's four Product Pillars to the OAT trips; we made sure the new trips delivered *unsurpassed value, good pacing, lots of choices, and the discovery of something new.*

Sure enough, within only 2 years, we had turned the company around, turning a $500,000 annual loss into a $3 million profit. That growth continued over the next 17 years: in 2011, more than 50,000 people traveled with OAT, and the brand brought in *$269 million in sales.*

We didn't have to buy OAT in 1993; other entrepreneurs might have felt they had enough on their plate with the acquisition of Grand Circle Travel only eight years earlier. But we saw the opportunity for this business to *complement* what we were already doing and expand our business. We learned that taking a longer view is a good idea, and so our advice is, don't get complacent in your business. Just because you're successful at what you're doing already, what *else* could you be doing that might enhance your business, make it more profitable? What else could you be doing to challenge yourself, your employees, and your company—while having some fun along the way?

Expand Your Business but Also
Stay Focused on Your Core Customers

Alan: In 1996, three years after we acquired Overseas Adventure Travel (and only one year after making it profitable), we decided it was time to get into a new category of travel: river cruises. Grand Circle had been chartering small ships for excursions on the Nile and Yangtze Rivers for years, but this new venture would expand on that limited experience in a big way. We jumped into the European river cruise business headfirst: first chartering a ship, then building our own, and then buying a fleet of barges.

What we did was risky, but our reasoning was sound: we recognized that river cruises offered a different kind of unforgettable experience that would appeal to older Americans. Most civilizations grew up first next to water, and so the oldest and most picturesque towns are frequently found along rivers and coastal waterways, making sightseeing easy. Small river boats and coastal cruisers can dock in shallow-water ports and at islands that can't accommodate large ships, getting travelers off the beaten path. Plus, the ships serve as floating hotels, and so travelers need to unpack only once, settle in, and move effortlessly between ports that promise new adventures every day.

We didn't know much about ships back then, but we knew how to deliver unforgettable experiences. Just as we had done when we acquired OAT, we leveraged our company's strong points, our Product Pillars—especially value and discovery. And we used our land expertise to design unique onshore excursions. Then we presented the trips to our travelers, who were clamoring for river cruises. In time, we were able to price our cruises on the assumption of 96 percent shipboard occupancy—far above the river cruise industry norm. This tipped the scales for us. Because most ship profit comes from selling the last cabins, we could offer our river cruises at prices far below those advertised by our competitors.

Today, Grand Circle Cruise Line is a world leader in small-ship cruises, operating more than 60 owned or chartered ships all around the world. Our ships ply the waters of Europe, South America, China, Egypt, and the Middle East. Our small ocean ships cruise the coastlines of Scandinavia, the Mediterranean, and South America. We also have deepwater cruises to Antarctica and the Galápagos. Of course, there's more to the story of how we got to this level of success on the water, but we'll cover that in Chapter 9.

Twenty-First-Century Marketing: Take a New Approach but Follow the Old Rules

Alan: Now that you know how we differentiate our Grand Circle and OAT brands and their products, let's take a look at how we market to customers. One of the biggest changes in our business since about 2005 or so has been in marketing. For more than 20 years, we operated as a direct-mail mass marketer, mailing tons of catalogs to our huge list of qualified U.S. households. Later, we figured out how to send very specific catalogs, by destination and by product type, to our list. This segmentation was very effective and was really appreciated by our customers, many of whom had complained about receiving too much mail from us.

With the growth of the Internet, our travelers began to have many options for learning about international travel and tours. We eventually discovered that our old way of direct-mail marketing was only one way to reach our customers—and that for many, maybe it was not the most effective. We needed to change the way we marketed.

The changeover from direct mail to targeted multichannel marketing was a big undertaking that affected many departments. Instead of dictating the change, we formed transformation teams of leaders and associates to tell us how to get it done. We really had no clear vision about what we were going to do; we just knew that mass marketing was no longer working.

It took more than a year, but the change was accomplished with little disruption to the business because it was guided by the people who had the most knowledge of the work and who had the most at stake. Today, we send highly targeted communications in many different formats, including letters, postcards, single-product brochures, Web discussions, and electronic news. We are able to target our communications because we know where our travelers have been and where they want to go next—because they let us know. Chapter 9, on thriving in change, tells more about the transformation process; the rest of this chapter describes our specific marketing techniques in more detail.

Customers Will Tell You What They Want

Harriet: To a surprising extent, travel is a trends business. "Hot" destinations come and go with world events, fluctuating currencies, popular fiction—even celebrity globe-trotting. In the 1990s, Eastern Europe and then China were hot; today's hot destinations include India, Southeast Asia, and South America. Many travel companies spend thousands of dollars each year trying to forecast trends. That makes no sense to us. We just ask our travelers, "Where do you want to go next?"

And they tell us—in their post-trip surveys, in letters, via our Web site, by phone, and by e-mail. And of course, they tell us every time they book a trip. More than 115,000 people traveled with us in 2011—that's a big database for spotting trends. Knowing our travelers' preferences helps us anticipate hot spots, build inventory in desirable locations, and offer trips we know our travelers want. Our travelers are our advisors, and we reward them by giving them unforgettable experiences in places they want to explore.

Send Targeted Materials—Not Mass Mailings

Alan: We are very skilled at really listening to our customers. When they tell us where they want to go next, we target our marketing and send them specialized catalogs that will interest them. This is part of our high-touch business practice, and it depends on having a relationship with our customers that is personal and information based. It is the kind of relationship that most customers want from us because it's what people want and expect from each other.

We didn't always take a comprehensive high-touch approach. In the early days of the company, our marketing approach was to acquire a list of likely travelers and then bombard them with catalogs, along with countless individual brochures and "personalized" letters. "Eastern Europe! South America! Prices slashed! Don't wait, Mrs. Johnson. Book today!"

This approach tended to polarize our customers. Some of them loved it: to them, every catalog was a catalog of dreams, and they pored over them, especially our award-winning Overseas Adventure Travel catalogs, which had big, beautiful pictures and lushly written itinerary descriptions. We even won a gold medal from the Direct Marketing Association for one of those catalogs in 1997.

But thousands of our customers hated all those mailings. We know this because we had thousands of complaints! "Take me off your mailing list!" "Mr. Lewis, don't you dare send me another catalog." "Think of all those trees!" One person even told us, "I'm using your catalogs to line my bird cage at this point!" We had built a relationship with these customers, but it was not the kind of relationship we wanted, and a breakup was looming. It took a while for us to understand that frequent, indiscriminate mass mailings were not the kind of high-touch relationship people wanted. They wanted to feel a personal relationship that responded to their own interests and desires, and they definitely did not want to receive so many catalogs.

We're not there yet: we still receive complaints about the volume of mail—but we finally get it, and we're working on it. For example, we recently launched a quarterly mailing plan for travelers who prefer that schedule, and we are relying more on e-mail for customers who tell us they prefer that way to communicate.

A U.S. Call Center Is Expensive, but It Provides a Superior High-Touch Customer Experience

Harriet: In addition to mailings, we also do an enormous amount of customer communication by phone. Our call center is not located in India or Ireland. In our call center on Congress Street in Boston, more than 200 associates answer as many as 11,000 calls a week from our travelers—calls to ask questions, calls to book trips, calls to change an itinerary, and calls to praise or complain. For most of our customers, their most personal contact with our Boston office is with that friendly voice at the other end of the line. When you call our main number, that voice belongs to Laverne Schaff, and it's live, not a recording. Laverne has been our phone and front-desk receptionist for 16 years, and she brings her hometown Texan charm, warmth, and humor to the building. She represents the spirit of personal connection that we want to offer every traveler.

We value that connection. When a customer calls, our associates have a lot of information about him or her as well as lots of information about our trips right at their computer, and so they can jump right in and help. Over the years, we have spent millions of dollars on technology for the call center to keep it up and humming. We also invest heavily in training, even sending our call center associates on trips periodically so they can talk about them from firsthand experience.

Business consultants always ask why we have our call center in downtown Boston. They tell us no cost-conscious company would put a call center in such a pricey place, and they invariably suggest that we move it offshore or to a low-wage region.

We're not fools; we've done the math. We know that outsourc ing our call center would probably save us $3 million a year. But those consultants don't understand our business model. We are *a high-touch company* with *a strong two-way relationship* with our customers. We want the call center in our own building so we can personally ensure that our travelers get the best possible service on the phones. In fact, every member of the leadership team listens in on live calls occasionally—it's called *double jacking*—so they can maintain direct contact with our travelers. Our call center leaders connect daily with the marketing department to coordinate, and so our message is consistent in print, on our Web pages, and on the phones. We keep the call center in the building so we can stay in direct communication with our travelers.

We don't mean to disparage companies that outsource their call centers, and if you decide that's right for your business, that's your decision. But in addition to calculating the *actual costs*, we think you should also consider the *benefit* to your customers of talking to someone who knows them, understands them (phoneti-cally and psychologically), and *listens* to them.

Relationship Building on the Web

Alan: As a company, we were "late adopters" of Internet technolo-gies, primarily because we—Harriet and I—are a bit technologi-cally challenged. (In the old days, I actually had a chart next to my PC that reminded me to "press 1 to turn on, 2 to get my e-mail," etc. Fortunately, I'm now very handy on my BlackBerry, iPod, iPad, and other devices!)

Embarrassing as it may be, by the time Grand Circle made it into cyberspace, many of our older customers were already there. That turned out to be a good thing, because once we made a big investment in our Web site, our traffic really took off. It was our children, Edward and Charlotte, who set us straight about the Internet. Cross-generational feedback is one of the benefits of a family business. We now appreciate the immediacy of e-mail communication, the ability to create personal spaces on the Web site, and the ease of exploring our trips online—all of which has made our communication with our customers so much richer. And the interactive nature of the site fosters the kind of two-way communication we're after.

For example, customers can now get trip information, find deals and discounts, and access their online accounts 24 hours a day. They can write reviews of trips they've taken and post them online for the benefit of other travelers (the reviews now number in the thousands—and we don't vet them). They can send an e-mail to Harriet from a link in "Harriet's Corner," a section of the Web site where Harriet and our travelers share stories and pictures from their travels. They can post a question, join a discussion forum, find a roommate for an upcoming trip, get a recipe, or sign up for an electronic newsletter—all with the click of a mouse. On our Grand Circle Foundation Web site (which we launched only in late 2010), they can learn about all our foundation projects; they can even make donations through the site. We love the way the Web makes our communication so easy, and we love the virtual community that gets bigger and better every day.

The Internet has also become very important to us in times of trouble, because it allows us to reach many customers quickly through e-mail and through updates and announcements posted on our Web site. This is how we kept our customers current on the news when the volcanic eruption in Iceland closed airports all over Europe in the spring of 2010, for example, and how we announced in March 2011 that we were headed back to Egypt following the revolution there.

Between the community building and the emergency messaging, our investment in the Web has already paid off. (We'll talk more about communicating during crises in Chapter 10.) Because we were so late getting started, it was a considerable investment. It cost us about $20 million in staffing, technology, and research over a few years. Then we tried to catch up so fast that our first Web site wasn't very good at all. We didn't plan well, and we made some bad decisions: the original site wasn't designed with our best travelers in mind. Then we tried to relaunch it, but that didn't work very well either, because we had no reporting about what was being done (and what wasn't) and also because the technical platform we used was outmoded. We realize we made some big mistakes in our Web presence; so we're starting over, and we've just launched our redesign.

We're not the first company to make mistakes on the Web; countless others have, and redesigns happen every day. The important point to keep in mind is that your Web site should not only promote your products or services but should also invite your customers to communicate directly with you. That open door is invaluable—as we've said throughout this book, the information we get from our customers is all the market research we need! Wouldn't *you* like to know more about what your customers really think about what you do?

The Best Customer Contact Is Always Face-to-Face

Alan: We foster our new and old relationships with customers through special events, like the parties we mentioned earlier and other get-togethers we arrange. These events are a lot of fun, with lots of travel talk. They are especially useful because they provide an opportunity for real-time, uncensored, two-way communication. Live events are the most effective tool for building the kind of customer loyalty that can endure across multiple products and last for many years.

Our live meetings with customers also help us connect with their needs. We're not big believers in professional focus groups (although we have done a few over the years). Instead, we believe a large audience and a traveling microphone give us better guidance than a few people sitting around a table talking to a marketing guy in a suit. We want more talkers and a more genuine exchange than a focus group can usually give.

In troubled times, we hold get-togethers more often so our customers can hear from us directly about our plans for dealing with the problems and about what special deals we may be offering. For example, after 9/11 and during the financial crisis of 2008 and 2009, it really buoyed us to meet with so many customers who were still eager to travel. At one of our post-9/11 events, we handed out small American flags to everyone—customers and Grand Circle associates alike. We were all still in shock from the recent events, but we waved those flags like crazy, somehow comforted and inspired by this simple act of patriotism and solidarity. We gave one another the confidence to keep on going.

In 2011, we hosted about 80 get-togethers across the country, averaging about 400 customers at each event. You can learn a lot from 32,000 customers when you meet with them in person!

In an interesting about-face, the virtual "Harriet's Corner" on our Web site has inspired a real-life, physical space on the garden level of our building at 347 Congress Street. This engaging room, which is also called Harriet's Corner, is open to visiting travelers and to our nonprofit community partners, who meet periodically to collaborate on joint projects. We also use Harriet's Corner for get-togethers and special events. The room offers an opportunity for us to share our long history and unique corporate culture with our travelers and their friends. We have pictures of our children and their art and photography. We have inspirational letters from our customers and from Grand Circle Foundation honorary board members, including one from the late Sir Edmund Hillary. We have pictures and stories from our team-building off-sites, and we have photos of our associates leading volunteer service events. We hear a lot from travelers during their visits to Harriet's Corner, and this inspiring new meeting space for them is another piece of our high-touch approach.

Of course, there is no better feedback than that from customers in the middle of a trip. They are excited and usually complimentary, but we're not on location to get a pat on the back. As we circulate among customers, we always ask, "What haven't you liked about this trip?" Sometimes we wince at what we hear, but these conversations are important for us, because, like it or not, our customers are always right. Think about what you could learn by talking directly to your customers—and then find ways to do exactly that.

Thriving in Change

Complacency can be fatal.
Keep moving. Keep changing.

Harriet: When we acquired Grand Circle Travel back in 1985, the business was pretty simple. Our traditional coach-based tours were purchased from ground operators and managed from Boston. Our associates were primarily American, as were our guides. And most of our trips went to Europe. We marketed directly to our customers through catalogs printed in huge press runs. And we made do with a workforce of about a dozen people.

Boy, have things changed! Today, our products include small-group adventure travel, river cruises, and small-ship coastal cruises, as well as the traditional Grand Circle Travel trips. Our trips are no longer traditional sightseeing tours, but rich, cross-cultural experiences that include home-hosted meals, school visits, and days visiting local communities. We now offer trips to all seven continents, including destinations we never dreamed of 26 years ago, like Russia, Vietnam, Tibet, Burma, Jordan, Namibia, and Mongolia. In fact, we've just begun offering philanthropy-oriented trips to Cuba through Grand Circle Foundation.

Our company has changed, too. Today, we have a worldwide organization with 34 offices in 31 countries, and more than 2,200 people work with us. We buy our trips directly from vendors in the host country wherever we can, eliminating ground operators and their commissions. We've switched from mass mailings to targeted multichannel marketing. We have belatedly embraced the Internet. In 26 years, we have transformed ourselves from a small, money-losing travel company into a global enterprise—and a leader in our industry.

We believe when something stops changing, it's dead. That includes businesses, nonprofits, government agencies, you name it. Grand Circle is always in flux. That's the way we want it, and that's the way it has to be, because international travel is a volatile industry. Complacency can be fatal. We have seen many travel companies fail because they thought their business model was bulletproof or because they just weren't paying attention. In this business (and many others), you must be ready to change at a moment's notice. This means more than staying open to suggestions. It means having a corporate culture that values speed, encourages risk taking, and thrives on change. It also means having a process ready to go, because change is inevitable.

We've already described in general terms some of the changes in our company, but this chapter focuses on how we *managed* those changes. We'll describe how we integrated the companies we acquired and where we went wrong with some acquisitions or failed to keep up with changing conditions. And we'll tell how we successfully expanded into a truly worldwide operation.

We hope our experiences will give you ideas about how you might expand your own company—and how you can manage change and growth successfully.

Acquiring a New Business—Successfully

Alan: We started thinking about expanding our company only a few years after we bought Grand Circle Travel in 1985. Just a few years later, in 1993, we bought Overseas Adventure Travel. We have always been attracted to adventurous trips ourselves: Harriet went to Africa with friends after college; I did a fair number of outdoor adventure activities in the 1980s; and together, we've shared adventures around the world. At the time, we were convinced that our customers would like adventure travel, too. They were curious, well-educated people; some had been to Europe, Africa, and Asia in the service; many had already done the traditional European sightseeing tours.

And then there were the baby boomers, coming up hard and fast—and in big numbers. The first wave of that famously independent-minded generation would hit 50 in 1996, and there were 70 million more right behind them. Our plan was to meet them with a wide range of trips in Europe and beyond. Acquiring an adventure travel company would broaden our reach both demographically and geographically, ensuring our longevity as a company and protecting it against regional travel disruptions.

We started searching out adventure travel companies right after we returned from Nepal in 1988. I took a hard look at Mountain Travel but ultimately walked away. Mountain Travel delivers excellent adventure trips, and it's terrific at promoting its products, which is what attracted me to it. But because it's based in California and we're in Boston, we finally decided it might be too much of a challenge to work with a company headquartered 3,000 miles away. And as we soon found out, even a company across the river from us would prove to be a challenging culture merge!

In 1991, travel in Europe and the Middle East came virtually to a standstill when the United States declared war on Iraq. The travel industry was in turmoil, and 40 percent of American tour operators went bankrupt. We were determined not to be the next casualty. We had two rounds of very painful layoffs, and we experimented with a number of domestic trips, but none of them were very successful. Clearly, we needed to find new destinations and new styles of travel that would appeal to our travelers.

Fortunately, we found our opportunity right across the Charles River, in an old, three-story house in Cambridge, Massachusetts. The company was Overseas Adventure Travel, which offered custom treks and small-group tours around the world, especially in East Africa, Nepal, Turkey, and Peru. OAT's associates were mostly in their thirties and forties, and they were passionate about travel. They reminded us of ourselves. That's why, in 1993, we bought the company.

The Challenges of Integrating a New Company

Harriet: OAT was a good match for us, but it wasn't a perfect fit. We loved the exotic locales and the small-group concept—10 to 16 travelers per trip. But OAT offered trips to Mount Everest and the High Alps, and these trips were too strenuous for our customers. Remember, our customers are aged 50 and older, and many are over 70. OAT trips were "high adventure." We were looking for something more like "soft adventure."

We were also alarmed by the balance sheet. The company was bringing in $4 million in annual sales, but it was losing about $500,000 a year. The trips were expensive; yet many were unprofitable. Some trips had only one or two departures a year and as few as 12 travelers.

However, we had seen both these problems before, in the trips we had inherited from Colonial Penn when we bought Grand Circle. So we weren't too worried. Our goal was to offer small-group adventure travel to older Americans at a more affordable price. We wanted our customers to experience the amazing cultural interactions that come when groups are small enough to get off the beaten path and relate one-on-one with people very different from themselves. We knew we could get there by rigorously applying our Grand Circle business practices. We would weed out the worst-performing trips. We would buy direct, which would lower the per diem cost. We would build on our four Product

Pillars (described in Chapter 6). And we would include international airfare in the package price.

Maybe we pushed too hard and too fast for these changes. The founder of OAT, Judi Wineland, moved on after a year, saying our vision for the company no longer resonated with hers. When she left, a few people left with her. We understood. We are vision-driven people, too, and we were sorry we couldn't see eye to eye. One of the greatest difficulties every business faces after a merger or an acquisition is integrating the people from the original company into the larger company. Some of the people who left were hard-core high-adventure enthusiasts. They were also young—in their twenties and thirties—and they didn't really want to be part of a company that was now offering softer adventures to customers in their fifties, sixties, seventies, and older. They also were passionate about their trips—and we admired their passion. But they didn't realize that about 75 percent of their trips were unprofitable.

So we focused on making our new company profitable. Sure enough, within two years, we were offering adventures priced $2,000 to $3,000 less than our closest competitors. Our trips also included international airfare (theirs didn't). More important, we had opened the world of small-group adventure travel to folks who'd never known it was possible to travel like this, and we took them to places they'd only dreamed of going. That was our vision: to help change people's lives.

In 1994, Overseas Adventure Travel officially joined the Grand Circle family. For a while, we ran it as a separate operation in Cambridge. But in 1999, we brought OAT's associates to Congress Street to work at Grand Circle headquarters. We had left it alone for the first five years or so because it was such a small part of our business in those days. Plus, it was running well on its own, and so we kept it in Cambridge. Then, we began to realize that OAT had significant growth potential—greater even than we had originally thought—and so we began to focus on this brand more closely. Also, we had quite efficient operations at Grand Circle, and we thought we would streamline our OAT and GCT operations (and costs) by combining them. Why support two mailrooms when you could have one? Why have two separate call centers? Two HR departments?

Focusing on OAT's growth and creating efficiencies were two of the reasons why we brought OAT over to Boston. But a third reason is that we wanted to be one company—we wanted our U.S.-based associates together. We wanted to build a team.

Unfortunately, the move was unpopular with some of the OAT staff. OAT had always had its own, off-the-beaten-track kind of culture, and there was another round of resignations. It had been a small company when we bought it, with only 22 employees, but when it became profitable, we had hired more people to run it. Still, of the original 22, only 3 were left (and those 3 are still with us today, 18 years later).

From this, we learned that managing an office across the river can be as hard as managing an office on the other side of the world. Culture is culture, and you have to be careful to respect it, not to step on people's toes. But combining cultures isn't always possible either, as we found out. The original OAT company culture wasn't as open as ours: not everyone knew what was happening in the company, as they do at Grand Circle. And the OAT folks didn't really embrace our unusual team-building exercises (described in Chapter 3). We tried our best to merge, but we accepted it when people opted out rather than trying to fit in.

We also learned we had to honor promises. OAT had long offered a "small-group guarantee"—not more than 16 travelers on any trip. It was a big part of OAT's value proposition. But after OAT moved to Congress Street in 2000, we started to make some exceptions; after all, larger groups are more profitable. Passenger counts on some departures crept up to 17, even 18. Our OAT customers let us have it. They called us. They wrote to us. They blasted us in the post-trip surveys. They let us know in no uncertain terms that this trend was unacceptable and that we needed to make our money some other way. At first, we basically ignored the issue. We had so much going on, and the difference between 16 and 17 or 18 passengers didn't seem like such a big deal to us. But it *was* a big deal to our new customers—and we had committed to a 16-passenger limit. So in 2004, we created an internal team charged with safeguarding this guarantee and controlling costs. And we've been true to our word ever since.

Today, Overseas Adventure Travel is the fastest-growing tour company in the United States. In the past 10 years, the passenger volume has more than doubled, to more than 50,000 travelers a year. Sales have more than tripled—to $269 million. In fact, most of our company's growth now comes from our OAT brand. What was once a sideline business for us accounted for 43 percent of our customers and 47 percent of our sales in 2011. Our OAT small ships have made *Condé Nast Traveler*'s Readers' Choice Awards and its Gold List since 2002. And *Travel + Leisure* magazine readers have named OAT as one of the world's Best Tour Operators for 8 out of 9 years in a row. When we returned from Nepal in 1988, we knew small-group adventure travel would change our company, but OAT's progress has exceeded all expectations.

Along the way, we learned a lot about managing change:

- *Keep changing.* If you don't, you're dead.

- *Stick to your values.* We always rely on open and courageous communication, risk taking, teamwork, speed, and quality—and, of course, the ability to thrive in change. Your values can get you through anything and will help you to move forward.

- *Focus on what you do well.* Focus on your Extreme Competitive Advantages and on what has worked for you in the past. We applied what worked at GCT to OAT, and it worked.

- *Don't be afraid to take risks.* You will make mistakes, but try, try again, reassess, and go full speed at what works!

It's important to note, though, that it remains very difficult to keep two of our brands, Grand Circle Travel and Overseas Adventure Travel, independent from each other. In 2010, we underwent a very significant and time-consuming audit of both brands to make sure we were keeping the brands separate in terms of their identity. Because our brands are housed at our headquarters in Boston, under the same roof, and because many of the same people work on both brands, the brands had morphed quite a bit. Today, they are separate, but we learned both how important it is for us to keep Grand Circle Travel and Overseas Adventure Travel clearly differentiated and how easy it is for them to lose their identities when housed in the same location.

Acquiring Another Business— This Time Unsuccessfully

Alan: In 1997, we were excited about our success with OAT and about starting another new venture, too: building our first custom-designed ship for our European river cruises (which we'll describe in the next section of this chapter). We were on fire, and so we decided to buy *another* company: Vermont Bicycle Tours, which we later renamed VBT Bicycling Vacations. Now we really had the Old World covered—by plane, train, foot, ship, bus, and 10-speed, soft-saddled, specialized bike.

This was a whole new world of travel we had never experienced before. That was the problem, of course. Bicycle enthusiasts are their own breed. They are generally younger than our GCT and OAT customers, for one thing, and they're more competitive. They wanted to cover a lot of ground every day, and they wanted to bring their own bikes. For almost 10 years, we tried to figure out this new business—and we tried and tried! We slowed down the pace of some of the trips. We offered an alternative to hilly Tuscany, for example, with a flatter section of the Italy tour. We even considered using motorized bikes as an option for travelers, but we realized the bikes were too heavy and the charge for the batteries lasted only for a few hours, and so they weren't really feasible.

In the end, we were unable to mesh VBT with our Boston-based travel companies. The VBT folks were a fun group. And they were generous: they sold used bikes to Grand Circle associates for only $50 apiece, and they even drove them down in a van to deliver them to their colleagues, who appreciated getting such expensive bikes at such a terrific discount.

But in the end, the VBT travelers and products were simply too different from ours. And while we were making headway, the company had become a distraction, taking us off focus from the core of our business, GCT and OAT; so we sold VBT in 2006. We broke even on the deal, but considering all the time and effort we spent trying to make it work, it was a big mistake.

Don't Let New Acquisitions Overshadow the "Old" Business

Alan: A few years after we acquired OAT and had made it successful, we noticed that sales on the Grand Circle Travel side of our business—i.e., our original business—remained relatively flat. The world seemed to be moving to small-group adventure travel. So we initiated a big transformational change to phase out the older Grand Circle Travel land trips in favor of OAT trips.

To bid farewell to Grand Circle Travel's land vacations, we held (another) "Irish wake" on the loading dock behind our building. We hired a hearse, the executive team dressed in mourning attire, and we even had a real coffin. Joe Cali, our executive vice president of marketing, gave the eulogy, and then associates filed by the coffin to pay their respects, dropping old GCT catalogs into the coffin before heading over to tables topped with kegs of Guinness and bottles of Jameson whiskey. It was a fond farewell to a brand we had loved for so many years.

But boy, had we make a big mistake! Within a month or so of our farewell to GCT, letters and phone calls came pouring in from loyal travelers who had been with us almost from the beginning and were beseeching us not to drop the GCT brand. Luckily, we have no problem reversing direction when we see we are on the wrong path. So we created a life-size, three-dimensional, stand-up representation of a Grand Circle customer that we nicknamed "Mrs. GCT," and we displayed her in our lobby with a sign reading "She's Baaaaack!" Then we reinstated our old GCT trips and even added some new departures. Reinvigorated sales told us this was the right move. Lessons learned: listen to your customers; reassess; admit mistakes; then react.

Expanding Operations in an Unexpected Way

Harriet: We bought OAT because we understood adventure travel, and we molded it carefully to fit our vision for Grand Circle's future. Our next wave of growth (no pun intended) was a different type of expansion altogether—the river cruise business. In fact, it was closer to how Alan originally got into the travel business: an opportunity came along, and we seized it. That's how we got into river cruising; and although it's profitable today, in the early days, we were in over our heads.

Here's how it all started: In October 1996, we ran a very successful special event, a Rhine River Valley cruise to celebrate the thirty-eighth anniversary of Grand Circle's founding. It sold out in a matter of days, and soon our customers were clamoring for more. This was a demand we hadn't really foreseen. We were eager to meet it, but there was a problem. The charter cruises available in Europe at that time were mostly downscale affairs. The only reason our anniversary cruise had been such a hit was because we had chartered one of the better ships, and we had customized every detail of the itinerary. We couldn't replicate that success on a broad scale. (We did try. Unfortunately, that trip garnered one of the worst "excellent" scores in our history—only 17 percent on a Russian river cruise in 1997.)

The biggest problem was the ships themselves. The river ships available for charter in Europe were dark, cramped, and eccentrically decorated, and they were in dire need of refurbishment. They really didn't meet American standards, and so we couldn't charter them for our travelers.

Then, in 1997, a golden opportunity fell in our laps. The best of the charter companies, a Dutch firm, was about to build a new ship. The keel had not yet been laid. We bought in on very favorable terms, becoming full partners and having the ship built to our specifications. Honey Streit-Reyes secured the contract and oversaw the build-out from our new office in Munich. The new ship, the *M/S River Symphony*, was a ship Americans would sail on. It had all outside cabins, air-conditioning, floor-to-ceiling windows in the dining room, bathrooms outfitted to ocean cruise standards, and balconies.

Over the next two years, we built two more ships with our Dutch partner, but it wasn't enough: demand for these trips continued to outpace inventory. We couldn't keep up. And although the ships were terrific, the onboard experience still wasn't up to our standards. We had to contract with local providers for food, service, entertainment, and housekeeping, and the results were erratic. Sometimes even the electricity didn't work. So we bought out our Dutch partner, built seven more ships, and got some outside help on the service side. We were in the ship business—big time. But we had to figure out a better way to manage these operations.

Learning by Trial and Error

Alan: It's hard to overstate how big a change the river cruise business was for Grand Circle. Our expertise was land tours. We knew nothing about shipyards, or docking rights, or housekeeping operations. But in the space of only five years, we owned all these multimillion-dollar ships! Between 1997 and 2002, we built 10 river ships, each with a passenger capacity of 120 to 164. We know an opportunity when we see it. And we're not afraid to take risks if we think it will get us closer to our dream.

Still, there we were, in the year 2000, trying to figure out how to guarantee our trademark "unforgettable experiences" and not take a bath. I took Mark Frevert and a team of senior executives from Boston to Spain for an off-site to brainstorm and set goals. Then we traveled to the new Munich office to get input from our buyers and program developers there. Among them was a Croatian associate who, by sheer coincidence, was a certified master mariner and experienced ship's captain. We thought that was a good omen.

We tested our early itineraries to see what our customers' responses would be. They liked the convenience of river cruising. After all, it is an easy way to get from place to place, and they had to unpack only once. They also liked the small number of passengers on the ship and the camaraderie that offered. And they enjoyed the opportunity to meet and engage with local people on shore. But they didn't want to sit on the ship all day. What worked on our land tours—visiting families in their homes, spending time at local schools, experiencing local cultures and ways of life—seemed to appeal to our river cruise travelers, too. We distinguished ourselves from our competitors by pairing the best ships in the business with the best onshore excursions. Other river cruise operators didn't have our expertise on land, and so we were able to deliver much better experiences. We offered winery tours, home-hosted meals, school visits, walking tours, and visits to local markets. We also divided each ship's passengers into several smaller groups, each with its own guide, to keep the experience up close and personal. So we were learning what worked, and we were improving our ship tours.

At the same time, though, while we were learning by trial and error, some of those errors turned out to be very expensive! For example, in late 1999, we were thinking about buying European Cruise Lines (ECL), a small river cruise operation, to expand our ship operations. As is customary in American business, our CFO first wrote up a letter of intent; he then flew to Holland to present the offer. In the United States, a letter of intent does not have the force of a contract. In Holland, it turns out, it does. Who knew? Not us. Suddenly, we owned a river cruise line!

There were four ships in ECL's fleet, and not one was up to our standards. We sold off two of them the first year. And we sank $8 million into refurbishing the others. But in the end, the venture was a bust. After a couple of years, we had sold all the ships and dismissed or reassigned all the 20 or so employees. All told, we had wasted $14 million, but we had learned a valuable lesson: *never sign anything in a foreign country without knowing what you're getting into.* We also realized we had given too much responsibility to our CFO at the time, who didn't know as much as he thought he did about international transactions. Whenever you're doing something new in your business, make sure you research thoroughly how to do it and make sure you involve someone with experience in that area.

Getting into ships was a crazy thing to do, really, and a lot of our advisors said we would live to regret it. It was a time in business when there was a lot of focus on core competencies—and ships sure weren't ours. We had to bring in a lot of expertise, and for a couple of years it was a fiasco. But we had a tiger by the tail, and we weren't about to let go. We knew, from our thirty-eighth anniversary European river cruise in 1996 that our travelers were eager to travel this way. Based on the high number of bookings we were getting, we knew they wanted to travel with us. We knew they wanted the cultural interactions they experienced on our land trips, and we set out to deliver these experiences. We knew that, over time and with some trial and error, we would figure out how to run a ship or two. So we weren't giving up.

How We Finally Turned the Ship Around

Harriet: Two initiatives finally nailed river cruises for us. First, we set up a yield management team in Boston to really push our load factors, i.e., the percentage of berths filled on each sailing. Our competitors' ships were running at only 70 to 80 percent occupancy. In contrast, because we had so many customers on our mailing list and because our marketing was so successful, we were able to fill 96 percent of our berths. That increase in the number of passengers on each ship allowed us to lower our selling prices. Soon, we were offering better cruises for far less money than was the competition—often more than a $1,000 less per person. We still have that price advantage.

Second, we developed an action plan to get more direct control of the shipboard experience. After several bad experiences with outside vendors, we began hiring crews and training the staff ourselves. We also established our own nautical and hospitality division, based in Dubrovnik, to deal with food, beverage, and housekeeping services. We now controlled the whole operation. Before long, we had complete vertical integration of the river cruise enterprise. This was something we had never attempted in any other aspect of our business. We became the only river cruise company catering exclusively to American travelers.

As we expected, once we had control of the entire operation, our quality scores soared. By 2001, our river cruises were off the charts. By 2004, we had 43 ships that we owned outright or chartered exclusively. Between 2007 and 2009, we took a new step, building three 4-star ships that would carry 50 passengers each for OAT. We had been chartering small ships and large yachts for OAT trips in the Mediterranean and the Galápagos for many years, but we had encountered the same problems of quality control that had plagued us on European rivers. Same problem, same solution: *build and outfit them ourselves.*

Today, we own or charter more than 60 small ships all around the world. Small-ship travel is a growth market for us. Our Grand Circle Cruise Line river cruises regularly receive awards and recognition in industry publications, such as *Condé Nast Traveler* and *Travel + Leisure*, including the World's Best Small Ship Cruise Line and Best Value Cruise Line. We always receive extremely high scores for the itineraries and onshore excursions that we offer our customers. This goes back to our Extreme Competitive Advantages— especially finding a niche market where we could offer something in a better way than could anyone else and where we could succeed and grow.

When we got started in ships, we were chasing an opportunity and running on instinct. We knew it was a big change for the company. Our execution wasn't perfect. But if we hadn't taken that risk, we would literally have missed the boat. What's important is to *seize an opportunity*. You need to take risks in order to grow. Choose clarity over certainty. You'll never know *everything* you need (or want) to know; so just made a decision and run with it. Bring your strengths to your new venture. Ours were to control everything ourselves and to deliver local, interactive, land experiences that could enhance the river cruises. Get started: test, try, make mistakes—but learn from them. Then, keep going!

The acquisition of OAT in 1993 and the venture into shipbuilding and ship operations in 1997 radically changed our company. Today, they are part of our niche market strategy, one of our Extreme Competitive Advantages that put us years ahead of our competition.

Building and Managing a Worldwide Operation

Harriet: In addition to growing our business by acquiring OAT and by getting into the river cruise business, we have also expanded our business globally. We have 34 offices in 31 countries.

We employ more than 2,200 people around the world. Business and philanthropy leaders often ask us how we manage this world-wide organization. They're puzzled by our success because they've stubbed their toes many times managing people from different cultures, even in such sophisticated business centers as London, Paris, Rome, Hong Kong, Singapore, and Buenos Aires. Our success in more exotic and far-flung places like Egypt, Vietnam, China, and Croatia leaves these executives scratching their heads. And they're less than thrilled when we tell them our secret: *just get out there and make mistakes.*

You really don't have much choice. To be successful today, you need to look beyond your own borders. The world is becoming smaller, and both businesses and philanthropic organizations must learn to raise their sights to the people who populate the entire globe. Grand Circle may have had more confidence going global: after all, we are an international travel company, and we love to take risks. But there is no magic recipe for dealing with other cultures. The encounter will be different for different industries and different corporate cultures. And the logistics are always difficult. It just takes experience. And experience comes from making mistakes. Here's how we've done it. It's not the only way to expand, but it has worked for us.

Opening International Offices Is Not an Easy Change

Alan: By 2003, we had opened up several offices overseas, and we began to see the tangible benefits of doing so. We were able to negotiate at a local level to obtain better rates for our customers. We had better control over our trips, and so our "excellent" ratings were up. We were there when problems occurred.

In 2004, we opened a record number of offices: 11 in one year—in Quito, Tokyo, Istanbul, the Galápagos, Phnom Penh, Buenos Aires, Copenhagen, Dijon, Budapest, Lima, and Letchworth, England. We also began to transfer responsibilities that were formerly assigned to Boston to our overseas offices. Besides buying trips and managing vendors, the regional offices were now responsible for all these functions:

- Training and managing guides

- Performing accounting functions

- Maintaining the computer records for their products

- Designing trip itineraries

- Preparing pre-trip information for travelers

- Handling any problems or dislocations that might occur on our trips

- Managing every aspect of our ship operations in their regions

At the same time, we were building new Grand Circle Foundation partnerships around the world. In 1996, we had started working with the Auschwitz-Birkenau Museum and the World Monuments Fund to support some of the world's great historic and cultural sites. As we rolled out our new overseas offices, we realized we could use the new organization to support more local projects, too, like village schools and orphanages. We saw we could also help develop projects directly with local communities, like the organic microfarm we support at the San Francisco School in northern Costa Rica. Our regional managers found new partnership opportunities for us. Our overseas associates helped us oversee the projects. And our local guides gave tours of the sites when our travelers stopped by to visit.

We felt this kind of radical decentralization was necessary. Keeping a tight rein from our Boston headquarters might have been easier—and certainly less nerve-racking! But it would have been counter to our vision, values, and corporate culture. Unless we devolved critical operations and responsibility on the overseas offices, we would never get leadership from the regions. Also, it didn't make sense for control of those operations to remain in Boston. We needed our overseas associates to assume ownership for the success of their products. If they didn't, we would lose our nimbleness in seizing business and philanthropic opportunities. And we would lose our responsiveness in crises.

People resist change, and our Boston office resisted the move to a worldwide organization pretty much from the beginning. There were logistical difficulties, like dealing with different languages and different time zones. But the discontent ran deeper than that. It was hard for our Boston associates to let go of the trips and responsibilities they had nurtured for so long. It was even more difficult for them to trust strangers who were thousands of miles away. Some associates who were terrific trip developers in their own right didn't have the interpersonal skills to support others in that job. Others didn't have the temperament; they were too competitive.

Our Boston associates dragged their feet, second-guessed, and withheld important information from the regional offices. They resisted trip design changes from overseas associates. They were slow to train them in our computer systems. And they put up barriers to transferring financial control.

Our Boston associates truly thought they knew best how to provide trips for American travelers. They didn't believe as firmly as we did that it's the locals who know best. And they didn't see that they were running the risk of cocooning our customers in an American envelope as they passed through every foreign country.

People especially resist change if they think it threatens their livelihood. In fact, the build-out of our overseas offices put few Boston jobs at risk. All the same, there was a certain amount of paranoia on Congress Street. After all, we had already fired all of our American tour guides and replaced them with local guides. How far could we be trusted with the jobs in Boston? Would all the work be transferred overseas? This was a very difficult period for us personally. We were at odds with some of our longtime employees. And we weren't getting the cooperation we needed to transform our organization.

The only thing to do was press ahead. We got our American associates out into the field overseas to train associates there. We brought regional people to Boston. We hired capable people in the regions to handle new tasks. We held off-sites on how to get the job done. And we communicated our goal and our reasoning over and over again. Grand Circle has a goal-oriented culture, and so setting hard deadlines for the rollout and holding individuals specifically accountable for making it happen probably had the most impact in turning the tide.

Eventually, people saw that we were growing so fast that saving jobs really wasn't the issue. Instead, getting help for the snowballing workload was the challenge. In 1998 alone, our associate ranks grew by 30 percent to more than 600 people worldwide. By 2000, we had 16 offices, and more than 2,000 people worked for us. Boston and overseas associates started to get to know and trust each other, and that also helped. It took longer than expected—several years, in fact. But momentum kept building until we were really moving at Grand Circle speed.

In typical Grand Circle fashion, we went a little too far, and we have since pulled some operations back to Boston. We have also changed some reporting lines. For example, certain overseas financial operations—finance, accounting, and payables—now report directly to Boston rather than through their regional managers. But we're still far more reliant on our worldwide associates than are other travel companies.

Harriet: Over the years, we have come up with a set of seven simple rules for running our worldwide organization. The logistics are more complicated, of course, and are somewhat particular to our company, but these seven rules give us guidance for many different situations, and we believe they can help other businesses as well. Here they are:

SEVEN RULES FOR RUNNING
A WORLDWIDE ORGANIZATION

1. Speak the same language.
2. Build on company values.
3. Communicate, communicate, communicate.
4. Set clear goals.
5. Impose financial controls.
6. Give back everywhere you go.
7. Respect people and cultures by treating everyone the same.

Rule 1. Speak the Same Language

Harriet: All our overseas associates must speak English. This applies to everyone from the regional manager to the accountants in the back room. We can have lots of differences, but language cannot be one of them. A common language is a prerequisite for effective global communication.

Even though everyone speaks English, we constantly remind ourselves that English is often not the first language of our associates. For this reason, we always ask for feedback on important communications to ensure that the message has been correctly understood. And we try to write our memos and updates in simple English, without jargon or slang.

Rule 2. Build on Company Values

Harriet: As in Boston, values count far more than work experience when we hire for an overseas position. We believe that if an associate shares our values, then we can teach the job skills. If the values are missing, even in an otherwise great résumé, we can never get that associate to become a full member of the Grand Circle family. We're not talking about cultural values here; we want our associates to hold on to the values of their national heritage and family upbringing. We're talking about the six company values we described in Chapter 2: open and courageous communication, risk taking, the ability to thrive in change, speed, quality, and teamwork. Adherence to a common corporate culture allows us to operate as a single, cohesive team, even though we come from very different national cultures with their own set of very specific values. Our company values are essentially the most important tool we use to run our global enterprise.

Rule 3. Communicate, Communicate, Communicate

Harriet: It is difficult to overestimate the importance of communication, especially face-to-face communication, in a global enterprise. At Grand Circle, we actually strive for overcommunication. We say, "You can lead from anywhere." But we can't get leadership from our overseas offices unless our overseas associates are absolutely clear on their assignments. So we hold a weekly teleconference with every office. We send Boston associates to help in the regional offices. And we bring overseas associates to Boston whenever we can.

The big event, of course, is our annual BusinessWorks. That's when the entire overseas leadership team, including all the regional managers, comes to the Grand Circle leadership center in Kensington, New Hampshire, for one to three weeks of strategic planning meetings. We deal face-to-face on the hot issues of each region, starting with the performance of the top people. Then we address the top issues with each of our top products. Regional teams share best practices with one another, resolve issues together, and plan for the upcoming year.

Written communications are also important. E-mail messages are traded constantly. If an overseas associate needs an answer from Boston, his or her primary contact in Boston tracks the query until it is resolved. Regional associates read our weekly electronic newsletter, *Bridges*, which provides a snapshot of the numbers, the hot issues, and the accomplishments and philanthropic activities throughout all the regions. Regional associates also approve all our catalogs, brochures, and Web site content to ensure we represent our trips correctly and don't promise more than they can deliver. At Grand Circle, communication is a two-way street, and one that is very well traveled.

Rule 4. Set Clear Goals

Harriet: We want consistently high-quality scores from our travelers. That requires a high-performing global team. We achieve high performance by making sure every associate has specific goals attached to his or her name at all times. These may be product excellence goals, inventory goals, cost goals, discovery goals, trip leader goals—whatever pertains to the associate's position. For example, a region might have a goal to improve excellence scores by 2 points. Or to acquire hotels and meals to accommodate another 1,000 travelers on a particular trip. Or to hold total costs flat over the prior year. Or to institute additional training for trip leaders who score below 80 percent in excellence on our travelers' post-trip evaluations. The goals will change from year to year, but the goals are always explicit and always tied to individual associates.

Our overseas associates live Grand Circle's values. They are dedicated and industrious workers, always on call to serve our travelers. Getting stuff done is not a concern; getting the *right* stuff done is the hard part. Clear and measurable goals get associates in Phnom Penh working in unison with associates in Cairo and Boston. And tending to goals is every associate's full-time job.

Rule 5. Impose Financial Controls

Harriet: Early on, we made the mistake of throwing too much out to the regional offices without instituting adequate financial controls. Large contracts are always tempting, and in some parts of the world, skimming a little off the top is considered standard business practice. Regional offices also handle a lot of cash for tips, entrance fees, and emergencies. The old adage that a few bad apples can spoil a barrel was certainly true in our case. We had one rogue associate in Munich who faked a list of vendors and issued checks to an Italian shell company for fabricated goods and services. Once a month, he would fly to Italy and transfer the funds to his own account. He was the most industrious of thieves, but we had a few more over the years, thanks to our lax controls!

After a few incidents of pilfering and outright embezzlement, we went against our natural aversion to paperwork and implemented strong financial controls in the overseas offices. Because a few unscrupulous people took advantage of our weak controls, every office now has to file monthly financial reports. And the books had better balance!

Rule 6. Give Back Everywhere You Go

Harriet: Grand Circle Foundation's motto is "Giving Back to the World We Travel." Giving back is good for the people we help, of course. But it's also good for business. The village innkeeper remembers that we support the local school when we come by to negotiate our contract for the next few years. Historic towns help us deliver unforgettable experiences because they know the foundation has made contributions to preserve local antiquities. When we search a country for people willing to provide home-hosted events, we get lots of applicants because people know we've helped their community through the foundation. Travel is always a symbiotic relationship. Giving back to the world we travel returns more than feel-good rewards. It makes it easier to lead a global enterprise.

We reinforce this reciprocal relationship by encouraging regional offices not only to support the foundation projects that our travelers visit in their countries, but also to engage in local community service. Volunteer service is not always the local cultural norm. In India, for example, it is customary for philanthropically minded people to donate money to a cause rather than do the work personally. So when Mohammed Iliayas, our general manager for India and Bhutan, organized a community service event in July 2009, the effort was all the more surprising for the recipients. In all, 44 volunteers, including OAT guides, staff, and vendors, spent a day renovating the Surdas School for the Blind in Agra, near the Taj Mahal. It was a nice example of putting corporate values to work to make a cultural difference.

Rule 7. Respect People and Cultures by Treating Everyone the Same

Harriet: Finally, leading a worldwide organization requires a genuine respect for other cultures. We share our six Grand Circle values with our associates worldwide. But we don't try to impose American culture onto other societies or individuals. Overseas associates wear what they want. They get paid time off for their national and religious holidays. They perform community service work that supports their cultural traditions. They decorate their offices as they please. And we never question their political convictions.

There is a business reason for this. Cultural disrespect makes people resentful, and a resentful associate is seldom productive. But there is a better reason to respect other cultures. Different cultures are interesting. And differences aside, we are all human beings. I've always said, "In every new place, no matter how different, I find people engaged in the same dance of life. In every person, regardless of language or dress, I see more commonality than difference. In travel, I discover true hope for global understanding." This attitude comes easily to us at Grand Circle because almost all our associates—wherever they work—are travelers. And real travelers approach the world with an open and inquisitive mind.

Avoid Two Common Mistakes of Going Global

Harriet: Building and leading a worldwide organization is difficult. It can be impossible if you make one of two common errors. The first is to treat your overseas offices like vendors—i.e., not extending leadership opportunities and providing parity, but just telling them what you want. This kind of relationship relegates overseas associates to second-class status, and they will notice *fast*. Our overseas associates have the same access to us as our Boston associates do. We share with them all the information that we disclose at our monthly corporate meeting. There are no closed books and no arm's-length dealings. If you do not want to treat your overseas associates as coworkers, extending them the same opportunities and responsibilities as your home associates, you would probably be better off contracting with an independent ground operator.

The second error is to manage overseas associates as if they were all Americans. We believe this is a lazy way to run a global enterprise—lazy and arrogant. People who grew up in Tennessee are different from people who grew up in Southern California. And Cambodians may have a hard time relating to either. We encourage all our associates to bring an open and fair mind to work. We also encourage them to do their homework, not just on the business operations in our overseas offices, but also on the business norms in our many far-flung regions. Misunderstandings are common in global enterprises. But cultural sensitivity can head some of them off.

Seven Simple Steps for Transforming an Organization

Alan: As you can see from all the changes we've made at our company over the past 26 years, there is no easy way to direct transformational change. It takes a clear understanding of what the change will mean at every level of the organization. It takes determination to make it happen. And it takes some careful tracking. We've messed it up more than once. But as we acquired other companies, as we branched out from land tours and soft-adventure programs to river cruises and small-ship adventures, and as we opened and delegated more to our international offices, we learned a lot about how to manage change. Just as we have seven rules for running a worldwide organization, we believe there are seven simple steps for transforming an organization:

SEVEN STEPS FOR CHANGE

1. Set clear goals and focus change where it will make the most difference.
2. Communicate the change at every opportunity.
3. Use off-sites to draw on the brainpower of the entire organization.
4. Get everyone on board.
5. Build transformation teams to direct the action.
6. Make actions happen: set deadlines and make sure you meet them!
7. Measure results and change course if you have to.

Step 1. Set Clear Goals and Focus Change Where It Will Make the Most Difference

Harriet: Look around you. Lots of things need to change, but you can't do it all. We believe it's crucial to pick the next most important change and then see it through to completion. A lot of companies fail at making transformational change because as soon as they get started on one change, they run off to fix something else. This is a formula for failure. An organization needs *sustained focus* to transform its business.

Step 2. Communicate the Change at Every Opportunity

Harriet: Communication is key. Leaders must formulate a clear goal and explanation for the change, one that everyone in the organization can understand. This is a challenge at Grand Circle, because our associates speak dozens of different languages. We must craft our message in simple but powerful English. Then we repeat it over and over again—in company updates, in our weekly e-newsletter, in conversation, in our weekly teleconferences with overseas managers, and in our monthly corporate meetings. It may get repetitive. Associates sometimes roll their eyes. But we have found that as soon as you quit talking about change, you lose momentum.

Step 3. Use Off-Sites to Draw on the Collective Brainpower of Everyone in Your Organization

Alan: The best brainpower seldom resides with senior management, and the best ideas don't always arise in the office. That's why we always start a major transformation with an off-site meeting dedicated to visioning the change. Put a group of smart, experienced people together in a new environment, preferably out of doors, and their solutions will never disappoint you. We reach deep into the organization to include people who know the day-to-day workings of the company and can see the land mines we might step on if we were to run willy-nilly through the landscape. Our off-sites don't always go as planned, but they *always* surface the hot issues.

Step 4. Get Everyone on Board

Alan: Over the years, we have developed a very structured approach to brainstorming and consensus building. We've had to do this because our company values encourage a kind of high-energy creativity that can sometimes go off in a thousand directions at once. The approach is based on a technique called the *nominal group process*, a decision-making practice that identifies key issues and considers all points of view, regardless of where they come from in the organization. We use techniques developed at our Grand Circle Leadership Center, which has been honing methodologies for this work for more than 20 years. Group facilitators keep the process moving forward in a speedy and predictable fashion. Because we have done this so many times, we know exactly what to do.

Step 5. Build Transformation Teams to Direct the Change

Harriet: Transformation teams are charged with making the change happen. Whether it's a new computer system or a new way to train trip guides, the process is the same. One member of the senior leadership team is identified as the sponsor and is held accountable for the initiative.

Together, the team members identify hot issues, determine necessary actions, set deadlines, and report progress to the executive team. Associates vie to get on these teams because we have a history of using them for testing and developing new leaders.

Step 6. Make Actions Happen: Set Deadlines and Make Sure You Meet Them!

Alan: The action plans are the important thing. At Grand Circle, action plans are always very specific. Every issue around a change generates a set of specific actions; each action is assigned to a specific person, who is responsible for getting it done by a specific target completion date. Since the goal is clear and everybody has participated in the plan, the transformation gets broad support, and every associate helps make it happen.

Step 7. Measure Results and Change Course If You Have To

Alan: Experience has taught us that when we are faced with a problem or a need for change, reacting quickly is more important than being exactly right. We can get away with fast action because we have an understanding with our associates that if an idea or action isn't working, we will quickly change our direction—without penalty to the team.

This practice has two prerequisites. First, our associates must feel free to say, "Hey, this isn't working." Second, we need to measure our actions to confirm progress. Measurement takes the emotion out of the assessment. It allows us to move forward in confidence and without blame. For this reason, all the actions in our action sets are measurable. Our metrics include such things as quality scores, sales figures, cost figures, and traveler volume. We constantly check the numbers to confirm that our change is on the right track. If we don't see what we expected, we reassess and make any necessary midcourse corrections.

Harriet: We didn't always follow these Seven Steps for Change. In fact, it took many years to come up with this list. We have learned a lot from the many transformations that we have chosen for the company. In the case of our river cruises, we didn't really know what we were getting into; we just saw the opportunity, took the risk, and developed the transformation plan on the fly. That's the hard way to do transformation, but it's not impossible, and it can be very rewarding—and very profitable.

There has been a lot of change in our lifetimes. Change will not only keep coming; it will continue to accelerate. As developing countries adopt modern technologies and market-based economies, they go faster and push us to go even faster to keep from being run over. Change can be stressful. It can also be exhilarating. Individuals can make a choice about how much change they want in their personal lives, but organizations must thrive on change if they want to survive and prosper. We may wish it were different, but no one can change the dynamics of a worldwide economy that keeps getting compressed in time and distance.

All this emphasis on change probably has something to do with our own personalities. Alan and I like change and challenges, and we seek out the road less traveled. We surround ourselves with people who have a similar mindset, too. We think it's the right attitude for our business and a major reason for our success. We don't manufacture semiconductors or automobiles; we deliver unforgettable experiences in every part of the world. When the world changes, we know we have to change, too. Some say you have to be a "change junkie" to work at Grand Circle. There's truth in that statement, and we say it with pride, too. For us, change is a constant—and a constant source of pleasure. We love travel, we love our business, and we love our customers. And we love watching our company constantly evolve into something new and unforgettable.

Alan: As chair and vice chair of the company, Harriet and I feel it is our responsibility to look outward and ahead. Most companies forget to look outside. Then, when a big change happens, they're blindsided and have to scramble to catch up. We strive to stay ahead of changes by spotting trends in their embryonic stage.

We do that by actively seeking information. We peruse dozens of competitors' brochures every week. We both read constantly, two or three books at a time—as well as lots of newspapers and magazines—and we keep up with the world political scene. We talk to our advisors all over the world and ask them what they see coming in the way of travel trends and business changes. We pay special attention to information from our own customers, who tell us what they think of our trips and where they want to go next. We read business journals. And we stay close to forward-thinking business professors, especially at Babson College, in Wellesley, Massachusetts, which houses our Lewis Initiative for Social Enterprise.

Last, but certainly not least, we travel. And wherever we travel, we listen to the local people and keep our eyes open to new opportunities.

Do we have perfect 20/20 vision? No. We've made tons of mistakes, but most of our mistakes have been in execution, not vision. Usually, we have thought so hard about a change that we get the direction right, even if we're not sure how to get there.

The changes we made to our company over the past 26 years were not minor. They were huge, and they came at a dizzying pace, sometimes with little warning. We don't mind that. We like change. In fact, people often remark that neither of us is very good at sitting still. And we know that change is a way of life in the travel business. That's why we made "thriving in change" one of our company values. Change happens quickly in the world, and a change can devastate a travel company—or any company—that is caught flat-footed. We always try to stay on our toes—and never on our heels—but it isn't easy.

These days, when we contemplate a major change, we focus hard on vision and planning. We know that if the goal is crystal clear, we will be all right. The path to get there may have a few unexpected detours, but that's OK with us. After all, we are travelers, and detours are sometimes the best part of the journey.

10

Managing Crises:
Moments of Truth

*Crisis is the norm in business today. Having a
process in place to manage crisis may help you not
only to survive it, but also to emerge as a stronger
organization.*

*A*lan: The international travel business is particularly subject to crises; we had known that from
our earliest days in the business. Other businesses may face the
occasional crisis—because of executive scandals, lawsuits, or re-
calls of defective products—but the international travel industry,
unfortunately, faces more crises more frequently than do other industries.

Our first year after buying Grand Circle was just one crisis
after another: airport attacks, the hijacking of the *Achille Lauro*
cruise ship, the crash of Air India Flight 182, the U.S. airstrike on
Libya, an earthquake in Mexico City, the meltdown at Chernobyl.
It was like an omen, but we were too busy scrambling to keep our
company afloat to see how essential it would be for us to become
expert crisis managers; that realization evolved over many years
and many crises. Over the last 26 years, we have come to see crises

as "moments of truth," times that test our leadership and engage our vision and values in new and difficult ways. We have also come to see that in crisis there is always opportunity, somewhere.

We once sat down and created a list of the crises we've had to deal with over the years, and they numbered around 300. The list is a true parade of horribles: accidents, natural disasters, an epidemic, wars, plane crashes, financial crises, riots, bombings, political demonstrations, and terrorist attacks. These are subjects that most companies (especially most travel companies) don't want to talk about, but we believe that looking at how to manage crises is helpful for people in any industry because *every* business will face a crisis eventually. The vast majority of our trips occur without incident, but when a crisis does arise, we know what to do, and we believe others can learn from our experience.

Here's just one example. The year 2010 was an unusually busy year for us—but not busy in the usual sense. That is, it wasn't busy just because we sent 115,000 customers off on more than 4,000 trips all over the world. No, 2010 was a particularly challenging year because there were so many natural disasters and so much political unrest that affected our trips.

One of the crises we encountered was caused by the volcanic eruptions in March and April 2010 in Iceland. Those eruptions spewed a cloud of ash that was 1,000 miles long, 700 miles wide, and 35,000 feet high—which is the cruising altitude for most transatlantic flights. As a result, airports in as many as 20 countries in Europe shut down, including those in Belgium, Denmark, England, Finland, France, Germany, Ireland, the Netherlands, Norway, Scotland, Sweden, Switzerland, and others. Those closures grounded 100,000 flights and stranded 10 million passengers for six days in April. In addition to stranding air travelers in Europe, people who had a connecting flight to Europe that originated in the United States, the Middle East, and Asia were also stuck wherever they were at the time. This was the highest level of air travel disruption since World War II—greater even than on

September 11, 2001, because that involved closure of air travel only over North America, whereas the 2010 volcano had farther-reaching effects.

That volcano affected thousands of our customers in Europe, India, North Africa, and the Middle East: we had customers who were on tours and scheduled to come home. Heathrow is a major hub for us, and so all those people were stranded. We had customers who were on ships in North Africa and the Middle East who were scheduled to fly into the airspace that had closed. And we had customers who were scheduled to leave on their trips but couldn't because no planes were flying.

So our crisis team met twice a day: first thing in the morning and then again at the end of the day. Yes, we have a crisis team: it includes our CEO and senior managers in marketing, customer service, and other operations; in the travel business (and maybe in *your* business, too), you need to have a crisis team that knows what to do. The team was constantly in touch with our offices in Amsterdam, Bratislava, Cairo, Dubrovnik, and Rome.

Our priority, as always, was to take care of our customers. We worried about the unexpected costs they would face as they waited to head home. We decided we would treat all our customers as though they had travel insurance, even those who didn't, so that no one would have to pay extra because a volcano had suddenly extended their vacation. We made arrangements for hotel rooms. We allowed our river customers to stay on their ships, or we put those customers into hotels and put new customers on the ships. We paid for meals. We paid for phone calls home so our customers could let their families know they were OK and so they could make whatever arrangements they needed to. We even arranged for additional tours wherever we could—at no cost—so that our customers wouldn't have to just sit around the airport or the ship or their hotel rooms waiting with nothing to do. We did this for the entire two weeks until the skies cleared and people could travel again. We absorbed about $2 million in costs that we didn't pass on to our customers.

It was a long, challenging crisis. But because we took the high road with our customers, we created a tremendous amount of goodwill among the people who travel with us—and we received positive press from the media—in contrast to a lot of negative publicity about other tour operators who made *their* customers pay for hotels, meals, and other expenses. We did what we thought was the right thing to do, and we were relieved that this crisis was over.

Volcanic eruptions may not affect your business, and you may not have thousands of customers stranded around the world for days at a time. But we guarantee your business will have *some* type of crisis, even if it's not an act of God. In our case, we knew what made us successful was paying attention to our customers, and so we needed to cater to them even during this crisis. You need to determine what makes *your* business successful and then amp up that feature during whatever crises your business faces; you need to do this whether you're a small business that makes specialty children's toys and you find out that one of your products is defective and potentially harmful, or you're part of a huge conglomerate dealing with an environmental accident and the accompanying negative publicity.

Thriving in Crisis Requires a Structured Response Plan

Harriet: Every crisis is different. But whether the crisis is a transportation failure, a terrorist attack, a financial collapse, a natural disaster, political unrest, or some other calamity we haven't yet imagined, we must be ready to handle it quickly. Over the years, we've developed an approach to crisis management that gets us moving fast. We don't dwell on external events because those are things we can't control. Instead, we focus on our mission and values and on our Extreme Competitive Advantages, because these have carried us through crises in the past. Our crisis response varies somewhat with the circumstances, but it tends to follow the eight actions listed below and described in the sections that follow. The important thing is to assess the situation, make a plan, and take action—quickly. If the fast plan doesn't work, we change course and try something else; we care less about the process than we do about the *results*.

THE GRAND CIRCLE CRISIS RESPONSE

1. Mobilize the organization.
2. Focus on actions that can make a difference.
3. Take care of people first.
4. Let customers decide what they want to do.
5. Plan, act quickly, reassess.
6. In a crisis, reduce prices.
7. In a crisis, find alternatives.
8. Above all, never give in.

Mobilize the Organization

Alan: The first thing we do in a crisis is get on the telephone to someone close to the problem in the field. It could be a trip guide, or a regional manager, or a ship's captain, or a vendor, or even an experienced customer. If no one is close, we send someone on an airplane immediately to conduct a firsthand assessment. The important thing is to communicate, communicate, communicate.

We are well positioned to do this because of our worldwide organization. With 34 offices in 31 countries, we have instant access to people on the ground, and they tell us what is really happening. It's like having our own wire service; in fact, we have more international offices than most news organizations. While the rest of the world is glued to CNN, listening to rumors and secondhand information, we are getting direct reports from our overseas associates, in English. We're also getting an immediate analysis of what the crisis means for our travelers and our business.

The on-the-ground analysis is the important part. It's what gets us off the mark fast. And it's only possible because of our corporate culture—when our values of thriving in change, open and courageous communication, risk taking, teamwork, speed, and quality are all urgently needed and go into overdrive, full throttle.

As we've said throughout this book, we believe locals know best. So over the years, we've turned over a great deal of responsibility to our regional offices. We've empowered them to contract with hotels, manage ships, design itineraries, hire guides, keep accounts, and deliver unforgettable experiences. Our overseas offices do not *represent* Grand Circle; they *are* Grand Circle. So when a crisis arises, we look to them for leadership and information. In fact, in an emergency, we will give them direct authority to make decisions that involve our customers.

For example, in 2010, political demonstrations in Bangkok erupted into violence and rioting. Bangkok is home to our largest overseas office with more than 50 associates and close to 100 guides, and Southeast Asia is one of our biggest markets. Our regional manager, Rung Chatchaloemuut, was trapped inside a hotel for 10 days, but we gave her full authority to decide how to manage the itineraries of our travelers in Bangkok. After all, she was in the thick of things. And she had a much better idea of how to keep the travelers out of harm's way than we did in Boston.

Also, time is one resource that none of us can ever get back. In ordinary times, it is the greatest business cost, but during a crisis, time can be your worst enemy. So you have to confront every crisis with the greatest possible speed. We do that by making hard decisions about the tough issues first. This is often about *people*, especially top people. We've found that the organization moves fastest in crisis when the naysayers and the fearful are relegated to the sidelines. So that's where we put them.

Finally, here's another important point. In a crisis, we do not retreat from our philanthropic involvement in the region. We continue to fund the schools, orphanages, museums, archaeological sites, and other initiatives that we support through Grand Circle Foundation. For example, when terrorists bombed the U.S. embassies in Tanzania and Kenya in 1998, we continued to support our schools in those countries. These projects are part of our business strategy—our travelers visit many of the sites—but they are also our friends and our partners. Supporting them is not a luxury of good times, but a part of who we are.

Focus on Actions That Can Make a Difference

Harriet: It's easy to get distracted during a crisis. Crazy things are happening. Without focus, people tend to flail around or retreat into unimportant details. This is true of almost everyone, including our own associates. We all need direction, fast, so we can take effective action. We know from long experience that we cannot control a crisis; we can only control our response to it. We also know that in a crisis, we must leverage the things we do best. In other words, effective crisis management requires a sharp focus on the few things that *can actually make a difference to the outcome.*

We gather the senior leadership and department heads and focus everyone's attention on our Extreme Competitive Advantages. Then we ask them to identify the top five issues using our leadership model, which focuses on top people, top vendors, and top products. Then we identify the top five actions to resolve those issues. These lists are put in rank order, so that we know what's most important and can deploy the top resources to the top issues. Keeping the leadership focused on our Extreme Competitive Advantages and business and leadership models is essential for achieving quick and decisive action in times of crisis.

With this fast flip-chart analysis, we have everyone's attention and the beginning of a plan. It is an application of the "80/20 rule," the management principle that tells us that 80 percent of our results will come from the top 20 percent of our resources and actions. We have never known it to fail. Later, this quick assessment will become a strategic plan.

Take Care of People First

Alan: Our first priority in every crisis is always to take care of our customers and associates in the regions affected. This might mean establishing direct communication with our trip guides. It might involve rerouting itineraries. We might need to arrange for hospital care, security, or embassy protection. Or we might need to book flights to bring customers back to the United States early. It also includes communicating with the designated relatives back home.

For example, when a Holland America bus crashed in Alaska in 1987, associates flew in from Boston to be with our hospitalized travelers. When an OAT van hit a horse-drawn cart in Tunisia in 1998, we mobilized our associates in the country to stay with the injured travelers. Also in 1998, when fires in Borneo closed airports in Malaysia, associates helped revise our OAT travelers' itineraries. The list goes on and on. We are experienced crisis interventionists; we know what to do for our travelers on the road.

The next priority is to provide as much information as we can gather to travelers scheduled to depart on upcoming trips to the regions. This includes news about current developments, safety information, and a list of options available if travelers choose not to take their scheduled departure. Of course, if a hot spot develops in a region that threatens our travelers, we will suspend trips to that area until the danger passes. For example, we did this after 9/11. We curtailed the development of several very promising trips to India and Nepal because of their proximity to suspected terrorist camps in Afghanistan. Although we were confident that these trips would be hugely popular in the future (as, in fact, they are today), we kept them off our books for almost three years.

Let Customers Decide What They Want to Do

Harriet: Our customers are adults with a lot of life experience. In a crisis, we believe we honor them best by explaining the situation and letting them decide for themselves what to do. For example, in 1997, 60 tourists were killed in a terrorist attack at the ruins in Luxor. We had 158 customers in Egypt at that time, and 113 of them were scheduled to visit Luxor the next day. We gave them a choice: they could skip the visit to the ruins, continue as planned, or return home. In the morning, 111 of our customers went to Luxor as planned. Of our 158 customers, only 2 went home, and that was at their children's insistence. Our customers trusted us, and they wouldn't allow terrorists to deny them the experience of a lifetime.

Plan, Act Quickly, Reassess

Alan: Once provisions are made for the customers, the senior leaders immediately put together a strategic response from the "top fives." The plan identifies goals, assigns roles, and determines the best process for getting the job done. We call this our "road map," and it is based in the GRPI model for team effectiveness, which is a tool we use that helps us define our goals, roles, process, and interpersonal relationships, whether as an organization, a department, or a team. It is not a detailed map. Instead, it's more like a bird's-eye view of how to get from where we are to where we want to be. Once the route is established, many hands will contribute to the details of planning and execution.

Most companies wait a week or more for the dust to settle after a crisis event. But we move fast and meet often: every two hours at the beginning, then at least daily until the crisis is resolved. To move quickly, we use what we call "directional information." Directional information is not all there is to know about a subject, but it's enough to set a direction. It's the information at hand, and every company has it. What is unique about Grand Circle is that we are not afraid to act on it.

We don't waste time. We don't wait for a committee report. We go full speed ahead toward the goal we have identified. We can do this without fear because we have no compunction about changing direction if the plan doesn't work. We say, "Act early. Act aggressively. Act often. Make mistakes. Reassess. Reassess. Reassess." We call this impulse for an immediate, focused response "clarity over certainty." And we consider it a key to good leadership.

Another unique feature of our crisis response is that we don't lock leaders up in a room by themselves. And we don't hire outside crisis consultants. We go directly to the people in the organization who can help. Just as we call on our overseas associates to get a clear picture of what is going on in the region, we call in every Boston associate who has firsthand experience with the current problem or who has prior experience handling similar crises. We will also tap people who we know will aggressively apply our values, especially open and courageous communication, risk taking, and speed. No matter their title, position, leadership experience, or length of service, they will be there in the conference room, helping the leadership team achieve clarity and direct the response. It is in times of crisis that our principle of Leadership from Anywhere really pays off.

Harriet: Having said that, there is one more aspect of crises that we want to discuss. *Crises are emotional events.* We all have feelings. Nobody is superhuman, certainly neither of us. We try to mentally prepare ourselves for dealing with tough situations—both inside the company and in the greater world. In troubling times, we support each other. We make a habit of assessing each other's reserves of courage and resilience. When one of us gets worn down, we see it as our job to remind our partner to take a break, with plenty of rest. No one can run full tilt every waking moment. We don't ask it of ourselves, and we don't expect it from our associates.

In fact, Alan is always telling people about his girlfriend SARA; he even has a doll with her name in his office. SARA stands for *shock, anger, rejection, acceptance*—the sequence of emotions that psychologists say most people go through in times of crisis. We've been down this emotional path many times. It seems that different people go through the sequence at different speeds, and sometimes they bounce back and forth among the stages. We've found that simply acknowledging the stages is a big help in getting through a crisis and back to more even ground.

We also remind each other to take care of our associates and never take them for granted. Crises drain everyone emotionally. We try to cushion the impact. We organize breaks and downtime. We find out whether family or friends have been directly affected by the crisis. And we send people home for rest when they seem to be at the breaking point.

Finally, when the crisis is over, we organize some crazy stuff so people can have fun. Everyone has earned it, and even if people don't feel like having fun, you need to do *something* to alleviate the strain. Then, after a couple days, we do an after-action report assessing what we did right and what we did wrong. As a company, and as individuals, we've always learned more from our mistakes than from our successes. We especially assess the performance of our top people and ourselves. A crisis forces everyone to live in vivid Technicolor, and so we take some notes while everything is fresh in our minds, knowing that the next crisis, unfortunately, is just around the corner.

In a Crisis, Reduce Prices

Alan: Many people stop traveling during a crisis. This has serious consequences for our business, and so we immediately lower our prices—often drastically. We can do this because as demand slackens or collapses, our vendors will offer us cost reductions. We simply pass these savings on to our customers. We've experienced this so many times that we actually cut our prices *before* getting concessions from our vendors. We trust them to help us, not only because we have built long-term relationships with them but also because we can help *them*. They've worked with us long enough to know that our customers will keep traveling, especially if the price is right, and so we can help fill their empty rooms, motor coaches, and restaurants.

Customers tell us over and over again that their most memorable trips are when everyone else stays home. Major tourist attractions are not crowded. And they can make wonderful connections with local people. At the same time, people in troubled destinations are grateful for visitors who help support the local economy, and they go to great lengths to make them feel welcome. It's a win-win situation: our customers get an unforgettable experience, and local people get our business and our compassionate presence.

In a Crisis, Find Alternatives

Alan: Since travel usually declines in troubled areas, we look for other destinations that will appeal to our customers. Then we offer great values on these alternative trips as well. This strategy provides our committed travelers with high-value alternatives, while attracting new bookings.

For example, when war broke out in Yugoslavia in 1991, we promoted our trips in Costa Rica. During the SARS epidemic in Asia, we promoted new trips to Canada. Today, we're offering great values and trips to South America and to Southeast Asia instead of promoting a lot in the Middle East due to uncertainty there. We can do this because we offer trips all around the world and have a worldwide operation that can accommodate sudden changes in demand.

The SARS epidemic in 2003 is also a good example of how important it is to trust your local associates, who typically have valuable information that others may not have. Early on, one of our associates in China sent us an e-mail, strongly recommending that we suspend our China trips until after the epidemic subsided. This was not what we wanted to hear since China was a big moneymaker for us. But it turned out the associate had good on the spot information, and we followed his advice. Our associate took a big risk that day. He had no special relationship with us in the Boston office. Also, his immediate supervisor disagreed with his recommendation. But he believed in our value of open and courageous communication. And we were grateful for his honesty.

Although we promote travel away from a region in crisis, we rarely abandon a trouble spot. We may suspend some departures or reroute some itineraries, but we keep our offices open and stand ready to accommodate any travelers who are willing to go. For example, we were the only U.S. tour operator to stay in Zimbabwe after violence there in 2000. That same year, 7,000 of our travelers visited Fiji despite a coup and rising racial tensions in the islands. Crisis in Israel? We stayed. Hoof-and-mouth disease in the United Kingdom? We stayed.

In March 2011, when Japan was hit with an 8.9-magnitude earthquake that triggered a massive tsunami, we didn't panic, even though we had 10 groups scheduled with more than 130 travelers in Japan. Instead, we immediately contacted our Tokyo office to find out what was really happening there, and we learned that in most of Japan, most businesses were operating as usual. We contacted our on-site guides, who spoke with our travelers, and of 130-plus, only 4 wanted to come home, which we arranged. We called family members on behalf of other travelers and arranged for any of our travelers to call home as well. Since the airport in Tokyo was open and since the travelers who were already there were moving forward with their trips, we continued to operate future scheduled departures to Japan (as well as any other trips that were scheduled to fly through the Tokyo airport). Finally, since the tsunami in Japan had initiated tsunami warnings along the South Asian Pacific coast, we contacted our regional offices in those areas and confirmed that all our travelers, trip leaders, and regional associates were safe.

Then we stayed in contact with all those offices and continued to monitor the ongoing disaster. On March 15, we canceled immediate departures to Japan until we could assess what the airport situation was and what was happening with the nuclear accident and its aftermath. The next day, we canceled all departures to Japan that were scheduled for the rest of March, and two days later, we canceled all departures to Japan that were scheduled from March through June. We also announced we would match up to $25,000 in donations, through Grand Circle Foundation, to support Japan's orphans. In less than 24 hours, we raised more than $60,000 from 615 of our travelers, *in addition to* our $25,000! By not panicking and by communicating constantly, we were able to ensure the safety of all our travelers who either were in Japan or were scheduled to go there, and we even were able to help some of the victims of this overwhelming tragedy.

Even after 9/11, we continued to travel to regions that other companies were avoiding, including both Egypt and Turkey. We had been to Egypt ourselves only the year before, and we remembered our guide, Farid, whose family had been in the tourism industry for five generations. Egypt depends on tourism, and the people have always been fantastic hosts for our travelers. We were concerned that Americans might now be reluctant to visit Egypt, a Muslim country, but if people quit traveling, how would these good people feed their families? Our whole way of thinking about the crisis was changed by our recent travels in Egypt. We wanted to help build goodwill between our countries, and so we decided that Grand Circle would continue promoting travel to Egypt.

In fact, on September 14, 2001, we sent our new ship, the *M/S River Anuket*, down the Nile. It was the first American-owned ship in Egypt, and it sailed only half full. But we persevered. In 2002, our Egyptian bookings were still down, but the travelers who went had truly unforgettable experiences. Our relationship with the Egyptian tourism industry, already strong because of our support of EgyptAir after its crash two years earlier, soon grew even stronger. That benefited our travelers and associates. We now have three ships on the Nile. While travel to Egypt declined in 2011 due to the revolution, we remain committed to returning there with travelers, full force, within the next few years.

Troubles come and go, and we want to be positioned to take lots of travelers to fascinating destinations when the crisis situation has passed. By continuing operations in a troubled region—or at least keeping our offices open—we maintain the trust and loyalty of our vendors, our foundation partners, government agencies, and our own associates in the region. Those relationships will pay big dividends over time. As Albert Einstein said, "In the midst of difficulty lies opportunity."

Above All, Never Give In

Harriet: A big part of our philosophy is to "keep traveling"—even if there are only a handful of travelers booked for a given departure. We want to be the travel company for those intrepid souls who will travel even when times are tough, even in times of crisis. We know that these people are our best customers. We learned that lesson after 9/11, when our most frequent travelers kept traveling and helped us weather the storm. We are so grateful to them—and we want to keep them as our loyal customers.

Passionate travelers will always find a way to go where they want to go. If they don't go with us, they will go with some other company—and we don't want that to happen. As we execute our crisis plan, we always have our eyes on the future, and that future requires that we keep traveling today.

Our Worst Crisis: The Crash of EgyptAir Flight 990

Harriet: The crash of EgyptAir Flight 990 was definitely our darkest hour, the most emotional of the many crises we have faced personally and as a company. It happened in 1999, and we were coming up on the new millennium riding a hot streak. Our river cruises were booming, and we had just opened five new overseas offices—in Thailand, China, Portugal, Italy, and Tanzania. Sales had just topped $315 million for the first time. Our biggest worry was that our computers would go haywire when the calendar rolled over to Y2K. Then we got the call, at 4 a.m. on a Sunday morning in late October, that a plane had crashed 60 miles south of Nantucket, just 50 miles from our Boston headquarters.

The call woke us both from a sound sleep. It was Mark Frevert, calling with all the information he had: that it was an EgyptAir flight, flying from Los Angeles to Cairo. It had stopped in New York, as scheduled, to pick up passengers and refuel. It took off late from Kennedy Airport and had flown uneventfully over Long Island and then over open water before suddenly, inexplicably, falling from the sky. There had been no SOS call, no radio communication at all with the tower, just a sudden, steep plunge into the waters off Nantucket. The Coast Guard was heading to the scene. There was no word of survivors.

We had 54 Grand Circle travelers on that flight: 42 for Grand Circle's Ancient Egypt & the Nile river cruise tour and another 12 for OAT's Cairo and the Eternal Nile trip. Nothing like this had ever happened to us before.

There was a lot of confusion about the cause of the crash. Was it mechanical failure? Terrorism? A deranged pilot suicide? To this day, we don't know for sure, and what became clear in the first several hours of the crisis was that it didn't really matter. Not to us and not to what we had to do. The investigation into causes and motives was outside our control. Remember, this is one of the ways we respond to crisis: we focus on what *we* can do, since everything else is out of our control. Speculation was pointless. We had work to do.

We needed to contact the travelers' families to break the news. We needed to take care of our travelers on the ground in Egypt and the ones scheduled to go later in the week. This is another way we respond to crises: we take care of our customers first. We also needed to meet immediately with our partners at EgyptAir. Alan was already on his way to our office on Congress Street; the phones would be ringing, and we couldn't let those anguished calls go unanswered. At the same time, as soon as dawn broke, I started making calls to associates, asking if they could come in to the office to help on the phones. Grand Circle is not a 24-hour operation, and this was not only Sunday, it was Halloween, and so I didn't know if we could get enough people to staff the call center.

I shouldn't have worried; 80 of our associates came in that day, many unasked, some in shock, some in tears. Many hadn't been on the phones in years, but they were willing to take any assignment; they just wanted to help. Crises bring out the best and the worst in people. On this occasion, we saw the best. Over a few short days, we devoted our hearts and minds to consolation, while grieving ourselves. No previous world event had ever hit us this hard personally, and none has since. We will be forever grateful to our wonderful associates who stepped forward in those days. Their kindness and compassion got us through.

So we mobilized our organization. Alan immediately sent members of the senior leadership to key crisis locations: two to Nantucket and two to Kennedy Airport, where the families were gathering and the National Transportation Safety Board had established a response team. Alan also traveled to New York later that day. In Boston, dozens of associates sat at the phones in the call center, talking to family members and arranging flights for them to New York. Another team was on the phones to Egypt, where we had 100 travelers already on trips. We offered them the opportunity to book early passage home or to change air carriers if they wanted that (none did); we offered them that choice because we believe in letting our customers decide what *they* want to do.

Our first priority was to take care of the families, and their shock and bewilderment was almost unbearable. For days, our devastated associates listened to their stories, fighting back their own tears as sons and daughters and grandchildren poured out their grief and confusion over the phone. Our associates had no special training in this kind of crisis; they would often finish a call and break down sobbing themselves, then sit back down to call another family member. We assigned one associate to each family, to give some continuity of care as we arranged for the families to come to Newport, Rhode Island, where the plane wreckage was being taken. Over the course of a couple of days, we spent more than $2 million to cover families' air travel and hotel costs.

More than 120 family members came to Rhode Island, 14 from one family alone. The National Transportation Safety Board was slow to make a decision about a memorial service, and so we arranged a private service for the Grand Circle families in Newport. We held our own memorial in Boston and threw flowers in Boston Harbor in memory of the victims. Later in the week, there was a public service for all the passengers with dozens of personal recollections of those who had died. The memorial services were very moving, but to be honest, we never felt much "closure." There can never be closure on so terrible a tragedy. We remember it all the time. But it did teach us that the best thing you can do in the face of death is *keep going* and lend a hand wherever you can.

The Terrorist Attacks of 9/11

Harriet: The crash of the EgyptAir plane was the most personal disaster we faced because we lost 54 of our travelers, but at least we knew how to mobilize and what to do in that type of crisis. The terrorist attacks on 9/11 were completely different, because all planes flying over the United States were grounded within an hour of the attacks and the airspace was completely shut down. Like the rest of the nation, we were stunned by what was happening. But we were also constrained from reacting as best we could because two of our strongest leaders—Alan and Mark Frevert—were stranded in Washington state, where they had been climbing in the Cascade Mountains. They wanted to get back to Boston right away, but there was no air travel, and so they were stuck on the other side of the country.

To make matters worse, the company's chief operating officer completely lost it: he couldn't make a single decision, and he soon left the office. Three weeks later, he left the company: that's not the type of leadership we need in times of crisis, or any time, for that matter. Not everyone can handle the stress of a crisis, however, not even designated leaders. We realized in hindsight that although this individual had a wonderful résumé, he had never fully embraced our values of open and courageous communication, speed, risk taking, and thriving in change—values that are critical to crisis management. Fortunately, one of our other senior people stepped into the power vacuum.

Alan: Canada was still flying, and so Mark and I rented a car and headed north. Mark drove, while I stayed on my cell phone with Harriet to deal with issues involving our customers and our associates. By the time we got to Vancouver, however, Canada had also shut down its airspace, and so we spent the night there and headed back to Seattle the next day.

On September 12, we were still stuck 2,500 miles from the office; it wasn't until Thursday that the air ban finally lifted, and we took off from Boeing Field in a chartered prop plane. The heat didn't work, and it took 9½ hours to get to Boston, but we were relieved to finally be headed home to deal with the crisis firsthand.

Back at the office, the two weeks following 9/11 were extremely emotional, with all of us struggling to get through them. The uncertainty of not knowing the fate of so many missing people was unbearable. People supported one another here in Boston, and the outpouring of support from the regional offices was overwhelming.

We didn't lose any travelers on that day. But our company suffered in the aftermath, because in the wake of the attacks, international travel came to a standstill. In October, our bookings were down 90 percent; by November, cancellations were up 80 percent. We had forecast sales of $500 million earlier in the year; we wouldn't even come close. The worst part of this loss was the effect on our people: we had to lay off 250 associates, 160 in Boston alone.

We weren't alone, of course. After 9/11, fully a third of U.S. travel companies merged or went out of business entirely. Fortunately, we had plenty of cash reserves and a sound recovery plan. Other travel companies looked for new business—and failed. Our approach was different. Instead of seeking new trips and new customers, we retrenched. In what would become our signature response to a business crisis, we focused hard on the things we do best.

We cut our product line by 20 percent, focusing on the trips with the highest quality ratings. We promoted our river cruises, the trips over which we had the most control, because we owned the ships. We retargeted our advertising, putting aside new prospects in favor of a list of 200,000 previous travelers, and we courted our loyal "Inner Circle" members. We renegotiated contracts with our overseas vendors so that we could cut costs, and then we passed the savings on to our travelers. We called this approach the Five Key Strategies, and we knew them by heart:

OUR FIVE KEY STRATEGIES AFTER 9/11

1. Cut underperforming trips to reduce product line by 20 percent
2. Focus on river cruises and our own ships
3. Target marketing to our best travelers
4. Build stronger relationships with Inner Circle members
5. Seek 25 percent cost reductions from vendors; then pass all savings on to travelers

The Five Key Strategies weren't guidelines or suggestions—they were law. Each Key Strategy had its own goals, and we measured every action and every outcome against them. When someone proposed an action, anyone within earshot felt free to challenge whether it supported one or more of the Five Key Strategies. For example, a radical new travel protection program passed the test: it allowed travelers to cancel a trip up to the moment of departure— something no other travel company had ever done before—which strengthened our relationship with our customers and gave them the confidence to book trips. Another radical suggestion, to cut all our all-inclusive programs, also passed the test: although the trips brought in $50 million in sales a year, their quality scores were relatively low overall. In this way, the Five Key Strategies disciplined the organization and got everyone moving in the same direction.

Adhering to our Five Key Strategies kept us focused and kept us going. It was the right strategy to take. But we also made many mistakes during this time, including being rigid with customers around our cancellation policies for those who called to cancel their trips. It took us close to two weeks before we decided to make our policy more lenient, but by then, we had already lost scores of loyal travelers for good, and we would later receive more than a dozen letters from attorneys general across the country about our initial cancellation policy. We had planned, acted, and reassessed. What we learned, which would serve us well in the future, is that being more lenient during critical times is essential to maintaining customer loyalty and goodwill.

It was a grinding climb back after 9/11, and we were often demoralized, especially after the layoffs, which came down in late September. But our travelers were a huge support to us during these challenging times. Seven weeks after the attacks, we invited about a hundred customers who had recently returned from Turkey, Egypt, and other hot spots in the Middle East to meet with us on Congress Street. The purpose was to get their advice on how to run the business in this new climate of terrorism. They also told us the best thing we could do is keep traveling. "Don't be afraid," they said. "Be the company that doesn't back down, the company that allows us to keep seeing the world."

Harriet: We took heart from the words of one traveler in particular, eight-time traveler Quinn Matthewson, of El Cajon, California, who wrote:

> *I have no doubt at all this country will prevail and that GCT and OAT will remain viable. You may temporarily lose a few clients, of course, but I would hope most sincerely, and believe, you will do well in the months ahead, difficult as it may be from time to time. I, for one, will continue to recommend people travel and do so with you. The reasons are the itineraries, the price, the pacing, the attention one gets, and your caring ways. You do have a lot of allies out there. You don't need my advice, I know, but this will be a long struggle for you. Keep the chin up. We'll all come out OK, stronger than ever.*

Our customers had such faith in us, and they gave us so much encouragement. We have a relatively young workforce at Grand Circle, but our customers are all 50-plus—and most of them well over 50. Their life experiences include the Great Depression, World War II, Korea, Vietnam, and the civil rights and women's rights movements. Their strength, perspective, and resiliency are amazing. Our mission is to help change people's lives, but on this occasion, they changed ours by giving us confidence and hope. We dug in, and in time, sales came back. Against all predictions, 2002 was a record year for the company, the most profitable in our history. We were thriving in change.

The Financial Crisis of 2008–2010

Alan: The most recent crisis to shake us up was the worldwide financial crisis that hit in the fall of 2008. We had seen recessions come and go, but nothing like this. Home values collapsed, financial institutions went into bankruptcy, automakers went into bailout, stocks collapsed overnight, and cash investments earned next to nothing. We had some alarming months in the fall of 2008 when our sales declined sharply because our customers were nervously guarding their bank accounts. We immediately undertook an aggressive cost-cutting campaign, with a goal of cutting $52 million from our 2009 contracts—savings that were then passed on to travelers in the form of steeply discounted prices. At an off-site in South Carolina in 2009, we identified other ways to help our customers travel. We instituted free domestic air travel. We waived most of the usual supplementary fees imposed on solo travelers. And we extended our usual 14-day risk-free guarantee to 30 days.

We put our best trip leaders on the maximum number of departures. We replaced some low-performing trip leaders. We launched "Harriet's Corner" on our Web site to communicate better with our travelers. And we offered a free matching service for solo travelers. Our mission was to leverage one of our Extreme Competitive Advantages—*unsurpassed value*. And we did it. Once again we came back strong; 2009 was the third-best year in our company's history, and our quality ratings were the highest ever.

Then, in April 2010, after an earthquake in Chile and mudslides in Peru, a volcano erupted in Iceland. It was almost comical: would there be no end to our tribulations? As we've already described, airports closed for six days all across Europe, stranding some of our travelers and keeping others from starting their trips on time. We got out of that mess, too, shortening some departures, allowing most travelers to travel at another time at no extra charge, covering the expenses of those who were stranded, and expediting our travelers' claims with the insurers. By the time the ash cleared and we could finally breathe a sigh of relief, there were brighter skies on the horizon: the dollar was at its highest point against the euro in four years, and our 2011 bookings were way ahead of projections.

Every Crisis Disguises Opportunity

Alan: One final point about crisis: there's a silver lining in every cloud, no matter how dark it may seem at the time. After EgyptAir Flight 990 crashed, it would have been easy to abandon the airline, as our competitors did. After all, the cause of the crash was unclear. But we work hard to establish our long-term vendor relationships. We knew the leadership of the airline personally, and we know the value of loyalty. Just as we don't leave a country when there's a problem, we don't abandon a vendor unless its agents do something dishonest or their quality no longer meets our standards. We continued to contract with EgyptAir at a time when the company greatly needed our business, and our loyalty has been repaid many times over the years in excellent air pricing, preferential flight availability, and concessions when we've needed them. We foresaw that opportunity and have profited from it.

A similar thing happened after 9/11. As bookings plummeted, our competitors canceled vendor contracts all over the world—flights, hotel rooms, ship berths, meals. We moved into the vacuum. We secured discounted long-term contracts with properties and services our competitors had previously locked up. When travel resumed, our competitors had to settle for second-tier properties at a higher cost. We emerged as the dominant player in several new regions, including South America and the Mediterranean.

Many opportunities occur because competitors freeze or panic. We know they won't slow down for very long, and so we move quickly to take advantage of their missteps. If we're fast enough, we can seize the opportunity while our competitors are still trying to come up with a plan. In doing so, we come up with a long-term advantage. Nailing opportunities requires nerve, speed, risk taking, and a long memory.

We've dealt with many crises since our first terrible year after buying Grand Circle. We know what to do. In fact, we now feel that if the company does not emerge from a crisis stronger than it was when we went in, then we have merely survived, not succeeded. Our goal for every crisis is to hang tough, leverage our Extreme Competitive Advantages, and emerge with a stronger leadership position.

Leaders Can Emerge from Crisis

Harriet: Some people rise to the occasion during a crisis. Others shrink into the kneewell of their desks. It has been heartwarming for us to see unexpected leaders emerge from troubled times, not only because they have given us hope and direction in the moment, but because they affirm our long-held belief that people can lead from anywhere.

In fact, nothing elevates leaders like crisis. We have seen it countless times. It seems that crisis can serve as a kind of leadership incubator, a place where personal growth emerges from the very act of facing adversity. People seldom feel comfortable taking the reins in time of crisis if they aren't also expected to lead in ordinary times. Crises can make leaders, but rarely outside of a corporate culture that empowers leadership every day and believes that people can lead from anywhere.

KEYS TO LEADERSHIP IN A CRISIS

1. A crisis is a gift:
   › Recognize that a crisis is a moment of truth.
   › Remember: in times of difficulty, there is always opportunity.
2. Your best preparation is your culture:
   › Develop a culture of leadership.
   › Stay true to your values: open and courageous communication and risk taking are critical.

(continued)

(continued)

3. If you do not emerge stronger, you have only survived, not succeeded:
   › Understand that you missed an opportunity.
4. Moments of truth are the best leadership test you will ever get:
   › Accept that some leaders will fail; move them out or out of the way.
   › Expect some leaders to excel; recognize and promote them fast.
   › Prepare for new leaders to emerge; move them up fast as well.
5. Act—be on your toes not on your heels:
   › Act early.
   › Act aggressively.
   › Act often—make mistakes.
6. Get a plan:
   › Move fast.
   › Be focused.
   › Get alignment.
   › Stay true to your core business—don't scattershoot.
7. Secure your leadership position—think three to five years ahead:
   › Identify your Extreme Competitive Advantages.
   › Seize the opportunity.
   › Keep in mind that the window of change is only open for a short time.
8. Overcommunicate:
   › Communicate the facts.
   › Communicate your plan.
   › Communicate the results.
   › Communicate the mistakes that were made.
9. Stop what doesn't work and go like crazy at what does:
   › Measure your results.

> Get everyone on the same page.
> Reassess and change course if needed based on results.
> Celebrate small victories and big wins.

10. Know your lessons for next time:

> Prepare, because there will be a next time.

Where We Give Back, 2011

Schools and Villages*

North and Central America
Costa Rica
San Francisco Primary School
San Francisco Village
San Josecito de Cutris Primary School
Sonafluca Primary School
Sonafluca Village

Guatemala
Oficial Parvulos School
San Luis de Pueblo Nuevo Village
Santa Catarina Barahona

Panama
General Manuel Benigno Higuero Guardia School
San Carlos

South America
Argentina
Escuela Rural Sol de Pacifico
La Concepción School

Chile
Chiloe
Quel Quel School

---

* Supported through Day in the Life program (in italics).

Ecuador
Sinamune Disabled Children's Orchestra

Peru
Chinchero
Las Palmas
Las Palmas Primary School
Pachar School
Pucruto School
Raqchi School
Urubamba
Villa Marcelo Primary School
Yanamono Clinic

Asia

China
De Ji Orphanage
Guang Ming Primary School
Hu Xian Donghan Village
Huo Kou Primary School
Shao Ping Dian Primary School

India
Adarsh Bal Vidhya Mandir
Khilichpur
Ramsingh Pura
Saini Adarsh Vidhya Mandir
Sri Venkateswara Orphanage

Laos
Baan Kia Luang School
Kia Luang Village

Nepal
Laxmi Primary School
Tomejhong

Thailand
Baan Boonyapark Early Childhood Center
Baan Mae Yang Rong School
Baan Pu Ong Ka School
Don Chum Primary School
Don Chum Village
Khon Sung Village
Machachulalongkom Buddhist University
Pu Ong Ka Village

Vietnam
Minh Tu Orphanage
Xom Gio Village

Europe

Croatia
Anton Tomaz Linhart Elementary School
Antun Masle School
Dobrisa Cesaric Elementary School
Gromaca
Kantrida Elementary School

Germany
State Museum of Auschwitz-Birkenau

Italy
Ancient Stabiae
C.F. Aprile Primary School
Corleone School
Santa Maria della Pieta School

Russia
St. Petersburg Music Boarding School

Slovakia
Gessayova Kindergarten
Haanova Kindergarten
Strecnianska Kindergarten

Turkey
Ahmet Ergun Primary School
Akburun Village
Akburun Village School
Alaattin Primary School
Ataturk Primary School
Belenbasi School
Belenbasi Village
Budak Village
Cumhuriyet Primary School
Dogancilar Village
Ephesus
Gobeller Primary School
Haci Bektas High School
Hacibektas Town
Hacibektas Veli Primary School
Hacibektas Vocational School for Girls
Ilicek Village
Kiriklar Village
Kiriklar Village Primary School
Kusluca Village

Middle East/Africa

Egypt
Bearat Village
El Taref School
Saad Zaghlol School
Valley of the Kings (World Heritage Site)

Israel
Lakia

Kenya
Amboseli Primary School
Amboseli Maasai Boma *Village*

Morocco
Dar ET-Taleb Education Center
Tineghir

Tanzania
Bashay Primary School
Banjika Secondary School
Kambi Ya Nyoka Primary School
Karatu
Nile Panda Primary School
Tarangire Maasai Boma *Village*
Tarangire Primary School

Zimbabwe
Ngamo Primary School
Ngamo Secondary School
Ngamo Village
Ziga Primary School
Ziga Village

South Pacific

Australia
Yiprinya School

Fiji
Sigatoka School

New Zealand
Kaitao Middle School

B

Recognition of Our Commitment to Give Back

- 2011–2006—*Boston Business Journal* cited Grand Circle among "Largest Corporate Charitable Contributors" in Greater Boston

- 2010—*Travel + Leisure*'s "Global Vision Award" for philanthropic travel

- 2009—West End House Boys & Girls Club "Passport to Belonging" Tribute

- 2006—Ernst & Young "New England Social Entrepreneur of the Year"

- 2006—Committee to Encourage Corporate Philanthropy "Excellence Award," with previous winners including Target, Whole Foods, and Timberland

- 2006—City on a Hill Charter School's "Citizenship Award"

- 2004—Neurofibromatosis "Cornerstone Award"

- 2004—The New England Women's "Leadership Award"

- 2004—Thompson Island Outward Bound's "North Star Award"

- 2003—UNESCO World Heritage Centre "First Partners in Conservation"

- 2002—AIDS Action "Corporate Partner Award"

- 2001—Big Sister Association "Achievement Award"

- 2001—Shelter, Inc.'s "Recognition Award"

- 2000—New England Chapter of the National Society of Fund-Raising Executives' "Hero of Philanthropy Award"

- 1998—Thompson Island "Founders' Award"

- 1996—Save the Harbor/Save the Bay "Founders' Award"

Sample Customer Survey

Full Survey

1 2 3 4 5 6 7 8 9 10 ...

CID 000000

Overseas Adventure Travel

Heart Of India (HOI, 04/29/2011)

Mr & Mrs John Smith
1000 Shelburne Lane
The Villages, FL 32162

Trip Evaluation

Trip Evaluation Instructions

This survey will be electronically scanned and processed which allows instant access to your ratings and comments by all of our worldwide associates. Your attention to the following instructions will ensure that your feedback is heard and acted upon.

* Rating Scale - Key
  E = Excellent G = Good F = Fair P = Poor
* Please use blue or black ball-point pen (not pencil or felt-tip marker)
* Fill in only one response per question and avoid creating stray marks
* Fill in circles as follows: ● ○ ○ ○   not like this: ✗ ✓ ⊙
* Print comments clearly and stay within the comment boxes. Comments outside of box cannot be reviewed.
* Please return ONLY THE SURVEY in the envelope provided. Please do not include letters, statements, travel incident reports, or other non-survey materials.
* Please do not staple the survey.

| Your trip evaluation begins here. |

1. Overall Rating of your travel experience

This section is used to evaluate the overall excellence of your travel experience. It is intended to measure your overall experience and is used as a baseline of all measurements. It is important for you to complete this section for results to be fully tabulated.

	E G F P
How would you rate the overall QUALITY of your travel experience?	●○○○
How would you rate the VALUE of your trip?	●○○○

	Yes No
Would you travel with OAT again?	● ○
Would you recommend OAT to others?	● ○

0 0 0 0 0 0 0 0 0 1

Page 1 of 12 - OAT Land 11v1

Index

A

AARP (American Association of Retired Persons), 5, 13
Accountability, 47–48, 82–83
Acquisitions, 162–169
Action, speed of (*see* Speed of action)
Action learning, 72–74
Action plans, managing change with, 188
Adventure travel, 149–150 (*see also* Overseas Adventure Travel [OAT])
Advisors (for charitable foundations), 80–82, 86
Alternative destinations, promoting, 204–207
American Association of Retired Persons (AARP), 5, 13
AmeriCares, 84
Apple Computers, 100, 135
Argentina, 85
Associates:
  asking for suggestions from, 62–65
  community service by, 96–98
  empowerment of, 19–20, 51–52, 76
  feedback from, 47–48, 133
  freedom for, 48–49
  hiring (*see* Hiring process)
  listening to, 125–126
  local, 110–115, 125–126
  questions from, 68–69
  at regional offices, 185
  resistance to change from, 177–179
  risk taking by, 146–147
  taking care of, in crises, 200
  as top priority, 51–53
  trusting your, 37–38
Associates' Fund, 96

B

Babson College, 30, 190
Bangkok riots (2010), 198
Barriers, breaking down, 70–72
Barth, Roland, 81
Bearat, Egypt, 93
Ben & Jerry's, 100
Bethune, Gordon, 105
Blow-in cards, 128
BMW, 100
Boston Family Shelter, 79
Brainstorming, 187–188
Brands, eliminating, 169–170
Bridges e-newsletter, 121, 181–182
BusinessWorks program, 70–75
Bush, George H. W., 29
Business(es):
  corporate culture as driver of, 30–31
  defining, with corporate culture, 34–36

Business(es) *(cont'd)*:
  new, 2, 4–5, 162–175
  "oxygen-mask" approach to, 10
  (See also Successful business-
    building)
Business practices:
  encouraging leadership with, 62–65
  high-touch, 135–137, 154–156
  low-touch, 135
BusinessWorks, 65, 181

C

CAG (Community Advisory Group),
  98–99
Cali, Joe, 58, 169
Call centers, 58–59, 155–156
Cancellation policies, 213
Carazo, Rodrigo, 87
Catalogs, 154
Chambulo, Willy, 63, 83, 87, 95
Change, 161–191
  acquiring new companies, 162–
    169
  expanding operations globally,
    175–185
  expanding operations with new
    businesses, 170–175
  maintaining focus during, 169–
    170
  seven steps for, 185–191
  thriving in, as corporate value,
    44–45
Charitable foundations, 80–82, 86
  (See also Grand Circle Founda-
    tion)
Chatchaloemuut, Rung, 198
Choice (Product Pillar), 106
Clarity, 7
Colonial Penn, 5–9, 12, 101, 104, 164
Communication:
  with associates, 16–17
  as corporate value, 42–43
  and crisis response, 197–199

Communication *(cont'd)*:
  with customers, 23–24, 121–122,
    126–127
  in Seven Steps for Change, 186
  at worldwide organizations, 181–
    182
Community Advisory Group (CAG),
  98–99
Community service, 95–98, 183–184
Community service teams (CSTs), 96
Company report cards, 48
Compensation, 131–133
Competitive advantage *(see* Extreme
    Competitive Advantages)
Complacency, 162
Confidence, 8–9
Conflict, 35–36
Consensus building, 187–188
Consultants, 125–127, 202
Continental Airlines, 105
Continuous leadership, 112
Core customers:
  building businesses with, 13–14
  identifying, 23
  knowing your, 137–138
  marketing to, 151–152
Corporate culture, 29–49
  defining your business with, 34–36
  as driver of business, 30–31
  effects of, 31–32
  freedom for associates in, 48–49
  and mission, 38–40
  philanthropy in, 78
  tone for, 32–34
  values in, 40–48
  and vision, 36–38
Corporate meetings, 67–69
Costa Rica, 87–88
Cost-cutting measures, 11–13, 215
Crises, 216–219
Crisis management, 193–219
  with crisis response plans, 196–208
  with cross-functional teams, 61–62

Crisis management *(cont'd)*:
  and EgyptAir Flight 990 crash,
    208–210
  in financial crisis of 2008-2010,
    214–215
  in international travel, 193–196
  and leadership in crises, 217–219
  and opportunities from crises,
    216–217
  and September 11 terrorist attacks,
    210–214
Crisis response plans, 196–208
Cross-functional teams, 61–62
CSTs (community service teams), 96
Culture:
  corporate *(see* Corporate culture)
  national, 175–176, 184, 185
"Culture of leaders," 63
Customer(s):
  communication with, 23–24, 121–
    122, 126–127
  as consultants, 125–127
  core *(see* Core customers)
  decision making by, 201
  expectations of, 132–133, 170–171
  face-to-face contact with, 126–
    127, 158–160
  involving, in philanthropy, 84–85
  knowing your, 137–138
  lifetime value of, 22–24
  listening to, 125–127, 140–141
  as partners in marketing, 153–154
  taking care of, in crises, 195–196,
    200
  understanding your, 142–143
Customer loyalty, 135–143
  benefits of, 138–139
  and Extreme Competitive Advan-
    tages, 22–24
  in high-touch businesses, 135–137
  and knowledge of customers, 137–
    138
  methods for improving, 140–142

Customer loyalty *(cont'd)*:
  and understanding of customers,
    142–143
Customer satisfaction, 131–133
Customer surveys:
  comments on, 46
  information from, 120–123
  as performance measurements,
    21–22, 119–125
  responding to, 123–125
  sample, 229
Customization (of service), 146–147

D
Day in the Life program, 92, 147–148
Deadlines, 188
Decentralization, 177–179
Decision making:
  by customers, 201
  in Leadership from Anywhere
    model, 53–54
  at newly acquired businesses,
    14–17
  taking risks in, 7
Developing countries, 114
Devine, Michelle, 121
Direct marketing, 102, 120, 128–
    129, 153
Direct TV, 141
Direction, 33
Directional information, 201–202
Disclosure, 68–69
Discount programs, 141
Discovery (Product Pillar), 107, 146
Discovery scores, 72
Dollar, purchasing power of, 83, 94
Don Chum, Thailand, 85
Double jacking, 156
Dreaming big, 5–6

E
ECL (European Cruise Lines), 173
Egypt, 206–207

EgyptAir Flight 990 crash, 62, 208–210, 216
80/20 rule, 199–200
Einstein, Albert, 207
Emergency messages, 157–158
Emerson, Ralph Waldo, 1
Emotions, crises and, 202–203
Employees, 19 (*See also* Associates)
Empowerment:
  of associates, 19–20, 51–52, 76
  and leadership, 66
  of regional offices, 198
Engagement, 75–76
Entrepreneurship, 2, 94–95, 98–99
Epstein, Bruce, 126
European Cruise Lines (ECL), 173
Excellence:
  achieving, 107–110, 123–124
  measuring for, 21–22 (*See also* Performance measurements)
Expansion of operations:
  global, 175–185
  with new businesses, 170–175
Expectations, customer, 132–133
Experiential learning, 59
Experimentation (by associates), 72, 131–132
Extended Vacations program, 104
Extreme Competitive Advantages, 13, 17–27, 124
  commitment to philanthropy, 20
  empowerment of associates, 19–20, 76
  focusing on, 17–19, 199–200
  lifetime value of customers, 22–24
  measuring for excellence, 21–22
  niche market opportunities, 24–27
  unsurpassed value, 20–21

F
Facebook, 122–123
Face-to-face contact, 94, 126–127, 158–160

Feedback, 14, 47–48, 133 (*See also* Customer surveys)
Financial control, 94, 118, 182–183
Financial crisis (2008-2010), 142–143, 159, 214–215
Five Key Strategies (after September 11 terrorist attacks), 212–213
Flexibility, 142
Focus:
  best products and services as, 13, 101–104
  during changes, 169–170, 186
  for competitive advantage, 17–19
  in crises, 199–200
  for philanthropy, 86–89
  when starting a business, 9
"Free passes," 72, 131–132
Frevert, Mark, 23, 33, 126–129, 172, 208, 211

G
Gates, Charlotte, 136
GCF (*see* Grand Circle Foundation)
GCT (*see* Grand Circle Travel)
Geopolitical threats, 9–10
GERT computer system, 11
Gestures, grand, 8–9
Global citizenship, 78
Global operations, 175–185
  common errors in, 185
  cultural differences in, 175–176
  opening international offices, 176–179
  rules for, 180–184
  value from, 113–118
Global philanthropy, 89–91
Goals:
  and mission statements, 38–40
  at off-site meetings, 71–72
  for philanthropic projects, 91–93
  in Seven Steps for Change, 186
  of worldwide organizations, 182
Google, 100

Grand Circle Foundation (GCF), 20,
    55, 59
  advisors for, 80–82
  beneficiaries of, 221–225
  and community service, 96
  customers' donations to, 100
  donations from, 99
  focus of, 86–87
  impact on customers, 84–85
  reciprocal responsibility at, 87–88
  recognition of, 227–228
  regional offices involvement with, 177
  response to Haitian earthquake
    by, 98
  tours of projects, 107, 147–148
Grand Circle Leadership Center, 30,
    74–75, 90
Grand Circle Cruise Line, 1, 152, 175
Grand Circle Travel (GCT):
  acquisition of, 5–9
  brands of, 1–2
  breadth of products, 18
  crisis response plan of, 196–208
  integration of OAT and, 164–168
  leadership model at, 52–54
  values of, 40–47
Grand gestures, 8–9
Ground operator system, 113–115
Group interviews, 55–57
GRPI model for effectiveness, 201
Guarantees, 166–167
Gutsy leaders, 66

H
Haitian earthquake (2010), 98
"Harriet's Corner," 122, 157, 159
"Harriet's Rules of Philanthropy,"
    89–95
Harrington, Barbara, 121
High-touch business practices, 135–
    137, 154–156
Hillary, Sir Edmund, 59, 81, 92, 93,
    160

Himalayan Trust, 81, 84
Hiring process:
  group interviews in, 55–57
  and hiring for values, 54–55
  for successful businesses, 10–11
  and unsuccessful hires, 57–58
  and value, 112–113
Hurricane Island Outward Bound
    School, 39

I
Iceland volcanic eruption (2010), 26,
    157, 194–196, 215
Iliayas, Mohammed, 184
Incentives, 130–131, 141
Independence, company culture of,
    32–33, 48–49
India, 184
Information:
  directional, 201–202
  providing customers with, 200,
    201
International travel industry:
  competition in, 24
  crisis management for, 193–196
  geopolitical threats in, 9–10
  guides in, 111, 112
  volatility/unpredictabilty of, 2,
    26, 53
Internet, 156–158
Interviews, 54–57
Invest in a Village initiative, 92, 93

J
Japanese earthquake (2011), 205–206

K
Kensington, New Hampshire, 90–91
Kibo Guides, 63, 83, 87

L
Language (for worldwide organiza-
    tion), 180

LEAD (Leadership, Exploration,
   Adventure, and Discovery) program, 70
Leaders:
   crisis management by, 202
   local, 91–93
   qualities of, 66–67
   unconventional, 62–65
Leadership, 51–76
   by associates, 20
   business practices that encourage,
      62–65
   challenges to, 42–43
   continuous, 112
   in corporate meetings, 67–69
   in crises, 217–219
   and engagement, 75–76
   entrepreneurial, 95
   at Grand Circle vs. traditional
      models, 52–54
   and hiring for values, 54–58
   Leadership from Anywhere model,
      53–54
   and learning, 58–61
   and off-site meetings, 69–75
   and qualities of leaders, 66–67
   reactive, 49
   of transformation teams, 61–62
Leadership, Exploration, Adventure,
   and Discovery (LEAD) program, 70
Leadership from Anywhere model,
   53–54
Learning, 58–61
   action, 72–74
   experiential, 59
   in teams, 60–61
   training for new employees, 58–59
   by trial and error, 171–173
Lewis Initiative for Social Enterprise,
   190
Lewis Institute for Social Innovation
   and Entrepreneurship, 30

Listening, 90–91, 125–127, 140–141
Local associates, 110–115, 125–126
Local involvement (in philanthropic
   projects), 87–88, 91–94
Local leaders, working with, 91–93
Local life, experiencing, 147–148
Location of business, changing,
   14–15
Low-touch business practices, 135
Loyalty (see Customer loyalty)

M
Managers, performance measurements for, 133
Marketing, 145–160
   and call centers, 155–156
   to core customers, 151–152
   and customer loyalty, 143
   customers as partners in, 153–154
   and differentiating your service,
      145–148
   direct, 102, 120, 128–129
   face-to-face contact, 158–160
   and improving your service, 148–
      151
   targeted materials for, 154–155
   twenty-first century approach to,
      152–153
   Web sites, 156–158
Mass mailings, 154–155
Matthewson, Quinn, 214
Meetings:
   corporate, 67–69
   with customers, 126–127, 158–
      160
   (See also Off-site meetings)
Microsoft, 77
Minh Tu Orphanage, 85
Mission statements, 33, 38–40
Mobilization (of organization), 197–
   199
Money channel, philanthropic, 94
Monteverde Conservation League, 84

Moody, Ralph, 140
Mountain Travel, 163

N

Napiteeng, Kipuloli, 59, 63–66
Neurofibromatosis, Inc., 78
New businesses:
  acquiring, 162–169
  expanding operations with, 170–175
  starting, 2, 4–5
  (*See also* Successful business-building)
New customers, acquiring, 138
New England Center for Homeless Veterans, 97
Niche markets, 24–27
Nominal group process, 74–75, 188

O

OAT (*see* Overseas Adventure Travel)
Off-site meetings, 69–75
  action learning at, 72–74
  breaking down barriers at, 70–72
  BuisnessWorks program, 70–75
  and Grand Circle Leadership Center, 74–75
  problem solving at, 70
  in Seven Steps for Change, 186
On-the-ground analysis, 197–199
Opportunities:
  being open to, 3–4
  and change, 175
  from crises, 45, 216–217
  niche market, 24–27
Organizational structures, 52–53
Outsourcing, 155–156
Overseas Adventure Travel (OAT), 116
  acquisition of, 162–164
  focus of, 24, 25, 149–150
  as GCT brand, 1
  integration of, 164–168

Overseas Adventure Travel (OAT) (*cont'd*):
  and reach of GCF, 82
  transformation teams at, 61
Overseas operations, 116–118 (*See also* Global operations)
"Oxygen-mask" approach to business, 10

P

Pacing (Product Pillar), 102, 106
Partnerships, 83–85, 153–154
Passion, following your, 4–5
Performance measurements, 119–133
  and compensation, 131–133
  customer surveys, 119–125
  and customers as consultants, 125–127
  in Extreme Competitive Advantages, 21–22
  for managers, 133
  for philanthropic projects, 82–83
  professional research surveys, 128–131
Persian Gulf War, 163
Personal involvement (in philanthropy), 89–91
Peterson, Maury, 81
Philanthropy, 77–100
  accountability in, 82–83
  advisors for charitable foundations, 80–82
  business benefits of, 99–100
  community service, 95–98
  as company commitment, 20, 78–80
  during crises, 199
  in Extreme Competitive Advantages, 20
  focus for, 86–89
  planning for, 83–85
  rules of, 89–95
  social entrepreneurship, 98–99

Philanthropy *(cont'd)*:
  at worldwide organizations, 183–184
Pinnacle, 58, 65
Price, 174
  in business strategy, 135–136
  in crisis response, 203–204
  and value, 105
Problem solving, 69–70, 74–75
Product Pillars, 102, 104–108, 124, 150
Products:
  differentiating your, 105–107
  eliminating, 125–126, 169–170
  focusing on best, 13, 101–104
  improving, based on customer feedback, 123–125
  profitability of, 103–104
  quality of, 103
  value of, 101–107
Professional research surveys, 128–131
Profitability, 103–104
Prybylo, Martha, 55, 59, 81

Q

Quality, 32, 45–46, 101, 174–175
Quality assurance team, 121–122
Questions (from associates), 68–69

R

Reactive leadership, 49
Reciprocal responsibility, 87–88, 94
Referrals, 138–139
Regan, Ricky, 67–68
Regional offices, 175–179, 185, 198
Repeat customers, 138–142
Report cards, company, 48
Respect, 140–141, 184
Results of change, measuring, 189
Rickert, Susan, 84–85
Risk taking, 7, 43, 146–147

Ritter, Charlie, 11
River cruises, 151–152, 170–175

S

Saga Holidays, 6–8
San Francisco School, 93
SARS epidemic (2003), 204–205
Sawyer Park, 91
Schaff, Laverne, 155
September 11 terrorist attacks:
  business opportunities after, 216
  crisis management after, 58, 62, 210–214
  meetings with customers after, 159
  suspension of trips after, 200
  travel protection after, 142
  travel to troubled regions after, 206–207
  and values of Grand Circle, 44–45
Service(s):
  differentiating your, 145–148
  focusing on best, 13
  improving your, 148–151
  local control, 110–115
Seven Steps for Change, 185–191
Shriners, 78
Simpson-Pye, Dine, 122
Sinya, Tanzania, 86–87
Small-group travel, 24–25, 149–150
Small-ship cruises, 25, 26, 151–152, 175
Social entrepreneurship, 94–95, 98–99
Social media, 122–123
Sonafluca School, 85
Speed of action:
  as corporate value, 46
  in crisis response, 198, 201–202
  when building business, 11–13
SquashBusters, 99
Starbucks, 136–137
State Museum of Auschwitz-Birkenau, 84, 177
"Stopping the bus," 146–147

Streit-Reyes, Honey, 37, 38, 115, 171
Successful business-building, 1–27
  by being open to opportunity, 3–4
  and competition, 27
  with core customers, 13–14
  by dreaming big, 5–6
  and early decision-making, 14–17
  with Extreme Competitive Advantages, 17–27
  by focusing on products and services, 13
  with grand gestures, 8–9
  and hiring process, 10–11
  with passion, 4–5
  and speed of action, 11–13
  and surviving the first year, 9–10
  and taking risks, 7
  when starting the business, 2
Surveys:
  customer, 21–22, 119–125
  professional research, 128–131
Sustainability, 92–93

T
Takeovers, 14–17
Tanzania, 84–87
Targeted marketing materials, 153–155
Teams:
  community service, 96
  cross-functional, 61–62
  learning in, 60–61
  quality assurance, 121–122
  transformation, 61–62, 188
Teamwork, 47
Thompson Island Outward Bound, 78
Timberland, 77
TNT (see Trans National Travel)
Tone, corporate culture and, 32–34
Training, 58–59, 61–62
Trans National Travel (TNT), 4–6, 9, 11
Transfers (of associates), 59
Transformation teams, 61–62, 188

Travel Companions program, 138
Travel protection, 142
Traveler appreciation events, 127
Trends, spotting, 153–154
Trial and error, learning by, 171–173
Trust, 37–38
Turnover, reducing, 56–57

U
"Unforgettable experiences," 37, 108–110, 115
United Travel Service, 3
Up & Out program, 97

V
Value (worth), 101–118
  and achieving excellence, 107–110
  of a company, 8
  and customer loyalty, 139, 141, 142
  as Extreme Competitive Advantage, 20–21
  from going direct, 110–113
  from going global, 113–118
  and hiring process, 112–113
  lifetime, of customers, 22–24
  as Product Pillar, 105
  of products, 101–107
  and profitability, 103–104
  in response to financial crisis, 215
  and value of products, 101–107
Values (ethical):
  actions based on, 47–48
  of Grand Circle, 40–47
  hiring for, 54–58
  importance of, 40–41
  of worldwide organizations, 180–181
Values statements, 33
VBT Bicycling Vacations, 168–169
Vermont Bicycle Tours, 168
Village-based discovery programs, 147–148

Vision:
  in corporate culture, 29–30
  creating your, 34–36
  inspiration from, 36–37
  and integration of new companies,
    165
  of leaders, 66
  philanthropy in, 79
  trusting in, 37–38
Volunteering, 79, 96–98, 183–184

W

Web sites, 122, 156–158
Weiler, Bob, 38–39

Wineland, Judi, 165
World Classroom initiative, 88–89,
  92
World Monuments Fund, 84,
  177
Worldwide organizations, rules for,
  180–184
Worldwide People & Culture department, 55
Written communication, 181–
  182

Z

Zaff, Greg, 99

About the Authors

Alan E. Lewis is owner and chairman of Grand Circle Corporation, the largest U.S. direct market tour operator of international vacations for older Americans. An entrepreneur, self-described maverick, and philanthropist, Lewis has a proven track record of creating both corporate profitability and growth and innovative philanthropic programs despite the challenges presented by a highly competitive industry operating in a chaotic world.

Harriet R. Lewis is owner and vice chair of Grand Circle Corporation and helps drive the company's mission to enhance the lives of its travelers, its global workforce, and the communities to which it travels. A former teacher, Harriet serves as chairman of Grand Circle Foundation, the organization's charitable arm, which is dedicated to supporting more than 100 schools and the communities in which they operate worldwide.

Grand Circle Foundation has, since 1992, donated or pledged $91 million to more than 300 humanitarian, educational, and cultural causes worldwide, including 100 schools in 60 villages in 30 countries.